T0137178

Advanced Sciences and Technologies for Security Applications

Indexed by SCOPUS

The series Advanced Sciences and Technologies for Security Applications comprises interdisciplinary research covering the theory, foundations and domain-specific topics pertaining to security. Publications within the series are peer-reviewed monographs and edited works in the areas of:

- biological and chemical threat recognition and detection (e.g., biosensors, aerosols, forensics)
- crisis and disaster management
- terrorism
- cyber security and secure information systems (e.g., encryption, optical and photonic systems)
- traditional and non-traditional security
- energy, food and resource security
- economic security and securitization (including associated infrastructures)
- transnational crime
- human security and health security
- social, political and psychological aspects of security
- recognition and identification (e.g., optical imaging, biometrics, authentication and verification)
- smart surveillance systems
- applications of theoretical frameworks and methodologies (e.g., grounded theory, complexity, network sciences, modelling and simulation)

Together, the high-quality contributions to this series provide a cross-disciplinary overview of forefront research endeavours aiming to make the world a safer place.

The editors encourage prospective authors to correspond with them in advance of submitting a manuscript. Submission of manuscripts should be made to the Editor-in-Chief or one of the Editors.

Theo Dimitrakos · Javier Lopez · Fabio Martinelli
Editors

Collaborative Approaches for Cyber Security in Cyber-Physical Systems

Editors
Theo Dimitrakos
School of Computing
University of Kent
Canterbury, UK

Javier Lopez
Department of Computer Science
University of Malaga
Málaga, Spain

Fabio Martinelli
National Research Council of Italy (CNR)
Pisa, Italy

ISSN 1613-5113 ISSN 2363-9466 (electronic)
Advanced Sciences and Technologies for Security Applications
ISBN 978-3-031-16090-5 ISBN 978-3-031-16088-2 (eBook)
https://doi.org/10.1007/978-3-031-16088-2

This Springer imprint is published by the registered company Springer Nature Switzerland AG
The registered company address is: Gewerbestrasse 11, 6330 Cham, Switzerland

Preface

Given the pervasively growing dependency of society on IT technology, and the corresponding proliferation of cyber threats, there is both an imperative need and opportunity to grow a new generation of young researchers trained to handle the changing nature of the upcoming cyber security challenges. These include evolvable threats and new technological means to exploit network and information security vulnerabilities of cyber-physical systems that have direct socio-technical, societal and economic consequences for Europe and the world.

As cyber security is inherently multi-sectorial, many dimensions must be considered and the new paradigms of cyber physical systems seem to be the most appealing. Indeed, cyber is becoming pervasive for many critical and user-centric application domains, and thus cyber protections will be paramount. We can envisage several application domains that would benefit cyber security research and development.

In addition, there is a significant advantage on sharing large amount of heterogeneous information in order to increase the capability of situational awareness, and the detection and prevention of advanced threats. In the past years, a number of security attacks with very serious consequences have been performed. These attacks have been successfully tackled thanks to collaboration among security and business companies. These companies shared relevant information, whose collaborative analysis has been vital in detecting the features to prevent and (automatically) respond to subsequent attacks of this kind. Indeed, there are several benefits in information sharing for cyber security (including incident notification) as well as several barriers to be removed.

In this book we collected a set of very relevant research results in the previous mentioned areas, often developed inside European research projects as NeCS, C3ISP, SPARTA, CS4EU and E-CORRIDOR.

The book will be indeed useful for researchers and practitioners in the field of cyber security with particular interest on cyber physical systems and collaborative approaches.

We would like to thank all the authors that contributed to this book and Springer team as well as our colleague Oleksii Osliak for contributing to several editorial aspects.

Canterbury, UK Theo Dimitrakos
Málaga, Spain Javier Lopez
Pisa, Italy Fabio Martinelli
June 2022

Contents

Acronyms

ABAC	Attribute-Based Access Control
ADP	Administration and Delegation Profile
AHP	Analytic Hierarchy Process
ALFA	Abbreviated Language For Authorization
AM	Attribute Manager
API	Application Programming Interface
AR	Attribute Retriever
ASIL	Automotive Safety Integrity Level
AT	Attribute Table
AUTOSAR	Automotive Open Systems Architecture
AWS	Amazon Web Services
BCEL	Byte Code Engineering Library
BSIF	Binarized Statistical Image Features
CA	Cyber Arena
CAN	Controller Area Network
CAN-FD	Controller Area Network Flexible Data-Rate
CbIP	Challenge-based Intrusion Prevention
CC	Contract Certificate
CCPA	California Consumer Privacy Act
CCS	Calculus of Communicating Systems
CDI	Compound Device Identifier
CH	Context Handler
CI	Consistency Index
CISO	Chief Information Security Officer
CMC	Charge Management Controller
CP	Charge Point
CPS	Cyber-Physical Systems
CPU	Central Processing Unit
CWB-NC	Concurrency WorkBench of the New Century
DAC	Discretionary Access Control
DBIR	Data Breach Investigations Report

DCB	DC Charging Box
DCT	Discrete Cosine Transform
DDG	Data Dependence Graph
DDS	Data Distribution Service
DFIR	Digital Forensic Investigation and Response
DICE	Device Identifier Composition Engine
DIDs	Decentralized Identifiers
DM	Decision-Maker
DMZ	Demilitarized Zone
DoS	Denial of Service
DP	Differential Privacy
DRM	Digital Rights Management
DS	Data Subject
E/E	Electrical/Electronic
ECSO	European Cyber Security Organisation
ECU	Electronic Control Unit
EDA	European Defence Agency
eIDAS	electronic Identification and Trust Services
eMAID	e-Mobility Account Identifier
ESP	Electronic Stability Program
ESSIF	European Self-Sovereign Identity Framework
EV	Electric Vehicle
EVA	Enhanced Vehicle Acoustics
EVCC	Electric Vehicle Communication Controller
EVSE	Electric Vehicle Supply Equipment
GDPR	General Data Protection Regulation
GRF	Ground Reaction Force
GSM	Global System for Mobile Communication
GW	Gateway
GXFS	Gaia-X Federation Services
HbIV	Host-based Integrity Verification
HE	Homomorphic Encryption
HIPAA	Health Insurance Portability and Accountability Act
HMAC	Keyed-Hash Message Authentication Code
HSM	Hardware Security Module
HV	High Voltage
IAM	Identity and Access Management
ICC	Inter-Component Communication
ICDFG	Inter-Component Data Flow Graph
ICT	Information and Communication Technology
IDS	Intrusion Detection System
IDSA	International Dataspaces Association
IDSM	Intrusion Detection System Manager
IETF	Internet Engineering Task Force
IoT	Internet of Things

IPSec	Internet Protocol Security
JSON	JavaScript Object Notation
KM	Key Management
LDP	Local Differential Privacy
LTE	Long-Term Evolution
MAC	Message Authentication Code
MCDA	Multi-Criteria Decision Analysis
MDP	Multiple Decision Profile
MFA	Multi-Factor Authentication
MFCC	Mel Frequency Cepstral Coefficients
MITM	Man-in-the-Middle
MITPC/Phone	Man-in-the-PC/Phone
MO	Mobility Operator
MPU	Memory Protection Unit
MSG BUS	Message Bus
MSYM	Multichannel Communication System
NbID	Network-based Intrusion Detection
NCR	National Cyber Range
NIST	National Institute of Standards and Technology
OCPP	Open Charge Point Protocol
OEM	Original Equipment Manufacturer
OIDC	OpenID Connect
OM	Obligation Manager
OMG	Object Management Group
OPA	Open Policy Language
OPC-UA	OPC Unified Architecture
OPC-UCON	OPC-UA Usage Control
OTA	Over-the-Air
OTP	One-Time-Passcodes
PAP	Policy Administration Point
PDP	Policy Decision Point
PEP	Policy Enforcement Point
PINs	Personal Identification Numbers
PIP	Policy Information Point
PnC	Plug and Charge
POM	Pairwise Ordination Method
Pub/Sub	Publish/Subscribe
PUF	Physical Unclonable Function
QI	Quasi-Identifier
RBAC	Role-Based Access Control
RGCE	Realistic Global Cyber Environment
RM	Reporting Manager
ROI	Region of Interest
RoTM	Root of Trust for Measurement
RoTR	Root of Trust for Reporting

RoTS	Root of Trust for Storage
RSU	Roadside Unit
RTT	Round-Trip-Time
SA	Sensitive Attributes
SCC	Secure CAN Communication
SCID	Sovereign Cyber-Intelligence Dataspace
SDLC	System Development Life Cycle
SD-WAN	Software-Defined Wide Area Network
SECC	Supply Equipment Communication Controller
SER	Security Event Reporting
SFA	Secure Feature Activation
SHE	Secure Hardware Extension
SM	Session Manager
SOC	Security Operation Center
SOTA	State of the Art
SSD	Secure Service Discovery
SSI	Self-Sovereign Identity
SSIs	Self-Sovereign Identities
SVM	Support Vector Machine
SWC	Software Component
SWIPO	Switching Cloud Providers and Porting Data
TCG	Trusted Computing Group
TCU	Telematic Control Unit
TEE	Trusted Execution Environment
TLEE	Trust Level Evaluation Engine
TLS	Transport Layer Security
ToE	Target of Evaluation
TPM	Trusted Platform Module
TPU	Tensor Processing Unit
UCON	Usage Control
UCS	Usage Control System
UCS+ nodes	Usage Control System Plus nodes
UCS+	Usage Control System Plus
UDS	Unique Device Secret
UML	Unified Modeling Language
V2X	Vehicle to Everything
VCs	Verifiable Credentials
VIN	Vehicle Identification Number
VPN	Virtual Private Network
XACML	eXtensible Access Control Markup Language
ZTA	Zero-Trust Architecture

Cyber Range Technical Federation: Case Flagship 1 Exercise

Tero Kokkonen, Tuomo Sipola, Jani Päijänen, and Juha Piispanen

Abstract Modern cyber domain is an extremely complex field to master. There are numerous capricious dependencies between networked systems and data. In cyber security, technology has a major role, but the knowledge and skills of the individuals combined with the incident response processes of the organisations are even more important assets. Those assets foster the cyber resilience of the organisation. The most effective ways to uphold these urgent assets are training and exercising. Cyber security exercises in particular have proven their efficiency in improving cyber security skillsets. During the cyber security exercises, it is possible to train cyber defence and incident response manoeuvres in stressful and hectic situations of being under cyber attack or intrusion. To achieve the capability to organise technical cyber security exercises with real attacks and real malware, technical training infrastructure mimicking real networks and systems is required. Such infrastructures are universally called cyber ranges or cyber arenas. Globally, cyber security exercises have become more common during the last decade, and there are several cyber ranges with diverse capabilities. Pooling and sharing the capabilities of cyber ranges raises the requirement to establish a cyber range technical federation. In this paper, a state-of-the-art implementation of the cyber range technical federation is introduced. In addition, the implementation demonstrated and evaluated during the Flagship 1 on-line cyber security exercise is discussed.

Keywords Cyber security · Cyber range · Cyber arena · Cyber security exercise · Technical federation

---

T. Kokkonen (✉) · T. Sipola · J. Päijänen · J. Piispanen
Institute of Information Technology, JAMK University of Applied Sciences, Jyväskylä, Finland
e-mail: tero.kokkonen@jamk.fi

T. Sipola
e-mail: tuomo.sipola@jamk.fi

J. Päijänen
e-mail: jani.paijanen@jamk.fi

J. Piispanen
e-mail: juha.piispanen@jamk.fi

T. Dimitrakos et al. (eds.), *Collaborative Approaches for Cyber Security in Cyber-Physical Systems*, Advanced Sciences and Technologies for Security Applications,
https://doi.org/10.1007/978-3-031-16088-2_1

1 Introduction

The modern cyber domain is extremely complex and includes complicated structures of networks and dynamic interactions of networked computer systems added with a increasing amount of potentially encrypted data. Understanding that entity requires special skills and awareness. Learning and experiencing during the cyber security exercises is an unquestionable fact. A well known quote attributed to general George S. Patton illustrates the fact quite aptly: "*You fight as you train.*" The importance of cyber security exercises is noticed in several national and international official documents. The EU's Cybersecurity Strategy for the Digital Decade [10] states that at the EU level awareness and exercises should enhance cyber defence capabilities and total cyber resilience, whereas Finland's Cyber Security Strategy [31] announces that both national and international exercises are utilised for ensuring the required high level education in the critical cyber competence. The importance of exercising is also noted in the proposal for a directive of the European Parliament and of the Council on the resilience of critical entities [11].

For achieving the capability to organise comprehensive cyber security exercises with modern vulnerabilities, attack vectors and malware, the total cyber domain shall be mimicked. For cyber security exercising, a cyber range can be understood parallel to a traditional shooting range that is serving competence to exercise skills with weapons, operations or tactics [36]. A cyber range shall implement a technical platform with the capability to simulate the required networks and systems for supporting the training & exercises (and also research & development activities) in the cyber domain. Cyber range is a centrally controlled environment including the required systems, tools and networks combined with a realistic Internet simulation, user simulation and background traffic generation. As a closed environment, cyber range offers risk-free usage of modern and realistic threat environments including real malware, attacks and intrusions [17, 23, 25, 26].

European Cyber Security Organisation (ECSO) defines a cyber range as follows [12]: "*A cyber range is a platform for the development, delivery and use of interactive simulation environments.*" They elaborate that a simulation environment represents organisation's ICT, OT, mobile and physical systems, applications and infrastructure. Such an environment could include simulation of attacks, users and their activities. Other simulated services, listed by them, could include Internet, public and third-party services. Furthermore, ECSO describes [12]: "*A cyber range includes a combination of core technologies for the realisation and use of the simulation environment and of additional components which are, in turn, desirable or required for achieving specific cyber range use cases.*"

Overall, there are many diverse cyber ranges implemented by security authorities, research centres, universities or industry all over the globe [37]. The requirements and perspective of cyber range development are often limited to specific use cases or capabilities, and therefore existing cyber ranges fluctuate from laboratory based test beds to tremendous virtual exercise arenas that mimic structures of the real global Internet. Cyber ranges are used for several objectives: as a platform for research, development and testing activities and also as an infrastructure for training and exercise transactions. For example, Deckard introduces a cyber-electromagnetic range executing kinetic and non-kinetic activities [5], while CyFRS is a fast recoverable cyber range based on a real environment [40]. He et al. introduce usage of cyber range for electricity grid as part of critical infrastructure [19], while Chen et al. reflect on the construction of a cyber range for personnel training in power information system [2]. Shangting and Quan discuss an industrial sector cyber range adopting QEMU-IOL virtualization technology [32]. Paper [6] proposed an approach for estimating the risk of compromise based on the data available from cyber ranges. Authors of [4] described the usage of cyber range for training the situational awareness in cyber defence. Authors of [38] reviewed in their position paper requirements for cyber ranges exploring the National Cyber Range (NCR) as a blueprint. Human-computer interaction, such as user interface, visualisation, design patterns and gamification, is also a concern when designing cyber ranges [33]. Lately, there have been activities to list existing cyber ranges. As a deliverable of Cyber Security for Europe project [3], the report on existing cyber ranges based on survey conducted across Europe has been released [34], and also the FORESIGHT project has produced a review of cyber ranges and test beds [37].

As illustrated above, the spectrum of cyber ranges and their usage are extremely heterogeneous. Karjalainen and Kokkonen introduce the concept of Cyber Arena (CA) for describing cyber range with the capability to simulate the total cyber domain including unexpected dependencies [23]. The different capabilities of cyber ranges have provoked the requirement for technical collaboration between different cyber ranges. Collaboration, pooling and sharing of capabilities in a cyber range federation enables an even wider compilation of cyber range capabilities.

In this paper, the implementation of cyber range technical federation is described and demonstrated during the Flagship 1 cyber security exercise, executed in January 2021 [21, 30]. The Flagship 1 exercise was executed as a remote, entirely on-line, exercise with participants from 22 affiliations from 15 different countries across Europe. The exercise showcased a cyber range technical federation using state-of-the art open-source Software-Defined Wide Area Network technology with great success. It is noticeable that compared to the on-site exercise, the on-line exercise has great challenges with exercise control functionalities, communication inside the Blue Team, communication between the Blue Teams and maintaining the situational awareness [24].

The rest of this paper is constructed as follows: the Sect. 2 describes the concept of cyber range federation. The Flagship 1 exercise with the description of technical requirements and implementation is illustrated in the Sect. 3. After that, in the Sect. 4, the results of participant questionnaire are analysed as the reliability assessment. Finally, results with found future research topics are concluded in the Sect. 5.

2 Cyber Range Federation

The basic requirement for a cyber range federation is quite obvious. The modern cyber domain is extremely complex, and different partners have different expertises. By the federation, those different expertises can be gathered and utilized. Several strategy papers indicate the requirement for co-operation in the cyber capabilities. For example, the National Cyber Strategy of the United States of America [35] indicates that partners have special cyber capabilities that can complement the existing ones. Also the EU's Cybersecurity Strategy for the Digital Decade [10] states the co-operation with international partners for strengthening the cyber defence capabilities.

Pooling & Sharing concept is noticed by both NATO and the EU. European Defence Agency (EDA) defines Pooling & Sharing as the EU concept which refers to increasing collaboration on military capabilities [13]. Smart Defence concept of NATO [27] includes Pooling & Sharing for generating cost-effective modern defence capabilities. A state-of-the-art example of Pooling & Sharing in the cyber domain is EDA's Cyber Defence Pooling & Sharing Project about Cyber Ranges Federation that showcased the technical co-operation and collaboration between different national cyber ranges at the European level [14–16].

The requirement for a cyber range federation is also noticed in the non-military domains. The EU launched four cyber security competence networking pilot projects [9], and all of those four projects have recognised the requirement for a cyber range federation. As stated by Graziano [18]: "*The idea behind a federation of cyber ranges is that multiple cyber ranges can be combined to provide greater simulation and scaling capabilities while leveraging on the vertical expertise of different cyber ranges.*"

The cyber range federation can be either operational or technical [34]: "*Operational Federation can be achieved "offline", without integrating or performing any technical federation of cyber ranges. The technical federation of cyber ranges enables the federated parties to utilize or consume specified functionalities, services, capabilities or resources from another party or parties of the federation.*" This paper focuses on the cyber range technical federation.

3 Case Flagship 1

The Flagship 1 exercise was conducted on 12–13 January 2021 as a unique on-line cyber security exercise available to partners of the CyberSec4Europe project [3, 21, 30]. The learning audience of the Flagship 1 exercise were from 22 affiliations across the Europe. During the exercise, the learning audience were placed into five different Blue Teams.

Blue Teams were simulating five individual Digital Forensic Investigation and Response (DFIR) teams of a fictional organisation known as University of Kybereo. The main task of DFIR teams was to investigate a response to a cyber security incident of the University of Kybereo. DFIR teams were using the provided incident response plans, communication guidelines, other documentation and required technical solutions. The roles of the learning audience were assigned based on the expertise queried prior to the Flagship 1 exercise [30].

The technical exercise environment was based on RGCE Cyber Arena [22]. Realistic Global Cyber Environment (RGCE) is a comprehensive cyber arena with substantial features. RGCE assembles in an isolated private cloud a realistic global world and real organization environments. RGCE implements thousands of virtual machines mimicking the global Internet and various organisation environments.

The objectives of the learning audience were to understand the DFIR process and team roles, technical investigation tools, communication within the team, within the organisation, its stakeholders and authorities. A hidden objective that was not exposed before or during the exercise was to understand the benefits of cyber security exercises for non security organisations. This objective was showcased to the learners by providing them incomplete incident response plans and communication guidelines. During the exercise they noticed incompleteness; however, they still had the main task to do, so the learning audience had to improvise. After the exercise, the provided documentation completeness was criticised. Actually, during the cyber security exercise an organisation may verify its processes, guidelines and documentation and after the exercise those can be further developed.

3.1 Technical Requirements of the Flagship 1 Cyber Range Federation

Technical specification and requirements of the implemented cyber range federation were based on the Piispanen's Master's Thesis [28]. In his thesis Piispanen introduces three different use cases for cyber range technical federation:

- **Networked cyber ranges** where different cyber ranges are connected to each other in a point-to-point, point-to-multipoint, or mesh-like manner for sharing the cyber range capabilities.
- **Extension of the cyber range's functionalities** where one cyber range serves as a provider (a hub) and offers connectivity to other cyber ranges that may provide additional functionalities to the hub.
- **Testbeds** where the cyber range offers domain-specific features such as testbeds or labs.

In the Flagship 1 exercises, the use case "*Testbeds*" was used. That use case is very similar to networked cyber ranges but in a smaller scale. The Master's Thesis did not cover the end user connectivity requirement and for this purpose, the remote end user connectivity use case was introduced in the Flagship 1 exercise. In addition to end user connectivity, the use case also specified the identification and the registration of users. The main requirements for the cyber range technical federation of the Flagship 1 were as follows [29, 34]:

- *"Specification 2.1: Overlay network SHALL support L3 connectivity into a cyber range (i.e. routed connectivity between cyber ranges)."*
- *"Specification 2.2: Overlay network SHOULD support L2 connectivity into a cyber range (i.e. extending L2 network between cyber ranges)."*
- *"Specification 2.3: Overlay interface SHALL support IPv4 and IPv6 connections in dual-stack."*
- *"Specification 2.4: Overlay network SHALL support IPv4 and IPv6 (cyber range Internetconnectivity does not need to be dual-stack)."*
- *"Specification 2.5: Overlay network SHALL support the following topologies: point-to-point, hub-and-spoke, partial-mesh and full-mesh."*
- *"Specification 2.6: Overlay network SHOULD support connectivity behind NAT/FW."*
- *"Specification 2.7: Overlay network endpoint SHOULD be implemented either in hardware or in virtual appliance."*
- *"Specification 2.8: End-to-End Round-Trip-Time (RTT) SHALL be less than 200 ms."*
- *"Specification 2.9: Overlay network SHALL have centralized management to control interconnections between cyber ranges."*
- *"Specification 2.10: Centralized management SHOULD be available to all cyber ranges."*
- *"Specification 2.11: Overlay network SHALL support segregation of concurrent exercises."*
- *"Specification 2.12: Overlay network SHALL be encrypted using industry standard protocols."*

3.2 Technical Implementation of the Flagship 1 Cyber Range Federation

The cyber range technical federation was demonstrated during the Flagship 1 exercise. The cyber range technical federation was based on open-source Software-Defined Wide Area Network (SD-WAN) technology. SD-WAN was chosen based on the features of security and flexible deployment options. SD-WAN allows configuration modifications during the execution and it doesn't require complete pre-configuration as for example site-to-site Internet Protocol Security (IPSec) or Virtual Private Network (VPN) solutions which are commonly used for federation purposes. The chosen open-source SD-WAN technology was ZeroTier, developed and open-sourced by ZeroTier Inc [39].

The capability for participants remote connectivity to the exercise environment was established by the technical federation. By the technical federation, also the features and functionalities of the RGCE Cyber Arena were extended by running a number of contents harmoniously in the Amazon AWS cloud [1]. The high-level cyber range technical federation environment is illustrated in the Fig. 1.

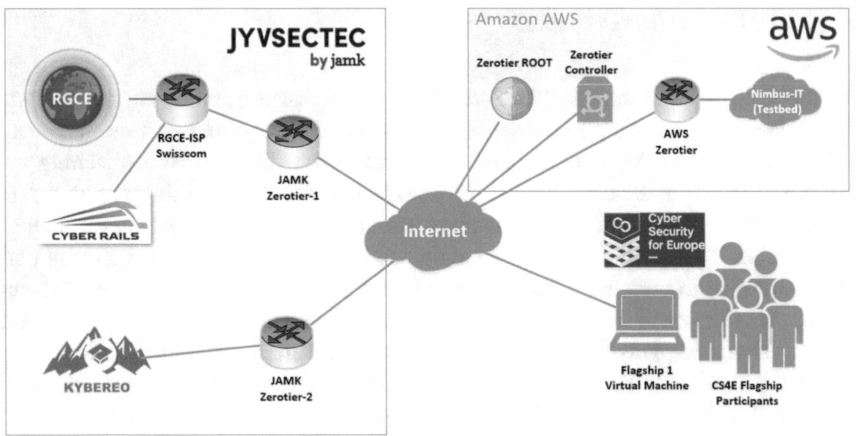

Fig. 1 Cyber range technical federation environment of the Flagship 1 exercise [29]

Cyber range technical federation environment of the Flagship 1 exercise included:

- ZeroTier SD-WAN infrastructure.
- Testbed, fictional cloud service provider that offered services to exercise organisations.
- Kybereon, Fictional University. The participants were part of the Kybereo's DFIR-team.
- Cyber rails, Fictional Train company. Partnership with Kybero.
- Swisscom and RGCE, Global internet functionalities and services.

The SD-WAN infrastructure of the technical federation was implemented on Amazon AWS cloud. ZeroTier Inc offers free public ZeroTier network, however a separated ZeroTier infrastructure on AWS was implemented. For the technical federation a SD-WAN edge router was deployed to AWS and to RGCE Cyber Arena. With these edge routers a secure network connection could be created between AWS and RGCE. A second edge router was also implemented. The second edge router was dedicated to remote users connectivity.

SD-WAN network controller was implemented because the infrastructure was disconnected from the Internet; thus the public Internet-connected ZeroTier's network controller could not be used. The developed controller used the JSON API of ZeroTier. Isolated networks can be created to the federation from that controller. The registration portal was implemented on top of our controller. When users was registered to our federation, the portal automatically added the end users as a member to their designated networks. A virtual machine image was created for the participant, which automatically connected to the federation network. The end users did not need to perform any configuration when they had the exercise's virtual machine, and the same virtual machine image was suitable for all participants because the configuration and network memberships were configured from our controller.

4 Reliability Assessment

A reliability assessment was conducted with a participant questionnaire during the final stages of the Flagship 1 exercise. The questionnaire was conducted anonymously without an indication of the respondent's identity or affiliation. When joining the exercise, the respondents were also informed about the usage of the provided data for scientific research and development of the environment in the privacy policy of the exercise. The questionnaire was conducted immediately after the technical part of the exercise to guarantee that the exercise audience had a strong emotion about it, and the experience was fresh in the participants' memories. As described earlier, there were participants from 22 affiliations across Europe. 21 individuals replied to the questionnaire. The questionnaire included closed yes/no questions and open questions for clarifying the feedback. Answers of the closed questions were analysed as quantitative data while answers of the open questions were processed by qualitative methods. First the answers of the open questions were coded (breakdown) [8] and then analysed by applying a qualitative content analysis [7, 20].

There were five Blue Teams in the exercise. The respondents of the questionnaire were distributed in the Blue Teams in accordance with Fig. 2.

The overall experience immediately after the Flagship 1 exercise was researched with the open question "*What are your feelings now?*". All the answers were positive without any criticism for the cyber range technical federation or conducted exercise. Some quotations from the answers illustrate the participants' feelings:

> "Very satisfied and excited. A really nice environment.", "Good, it was great learning experience.", "Superb platform, interesting scenario, good guidance.", "I found the exercise really interesting."

There was also a closed yes/no question to find out whether the exercise was beneficial for the participants. All of the respondents indicated that the exercise was beneficial, which was clarified with the open question "*Please describe how?*".

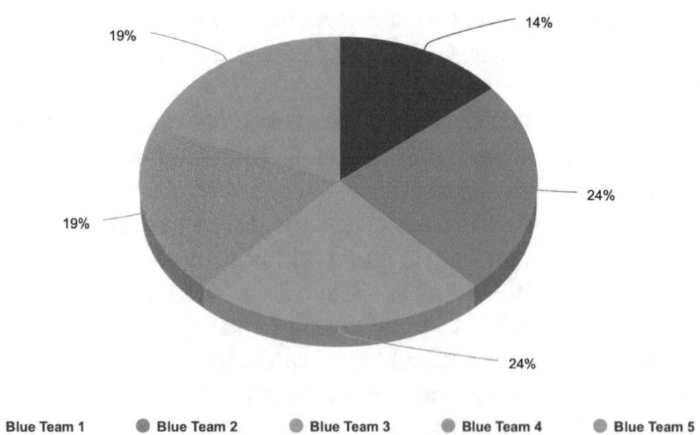

Fig. 2 Distribution of participants in Blue Teams

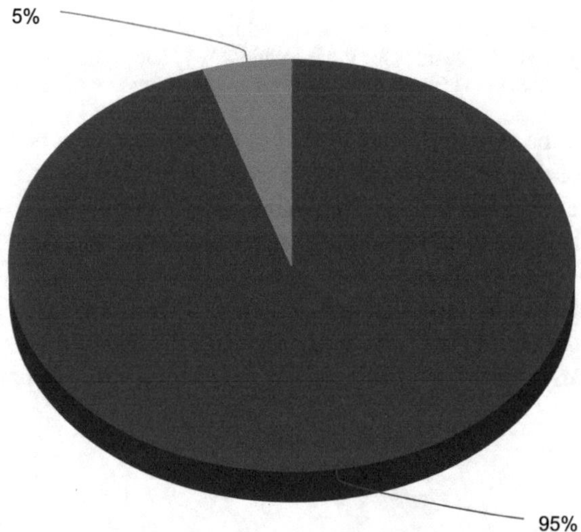

Fig. 3 Answers to the question "Did you learn something new?"

Noticeable is that even though the exercise was conducted on-line, also the teamwork was indicated there. As an illustration, some quotations from the answers are listed as follows:

> "It's a great way to learn.", "To learn the overall methodology and using new tools.", "Now I have idea on how these exercises really work.", "New tools and techniques + teamwork." "I was exposed to a very good cyber range."

It was also queried with the closed yes/no question whether the participants learned something new during the Flagship 1 exercise; this was further clarified with the open question "*Please describe what?*". As illustrated in Fig. 3, one of the respondents (5%) indicated that they did not learn anything new, but all the others (95%) indicated that they had learned something new during the exercise. It is noticeable that there were no critical open comments about learning something new. Following quotations from the answers illustrate this:

> "Learned a lot. The teamwork was perfect.", "Procedures, methodologies and teamwork around cyber security.", "Use of new tools."

It was also investigated whether the participants would recommend such an exercise for their colleagues. All the answerers indicated positive recommendation of the exercise. Finally, there was an opportunity to give open feedback. Most of it was positive; however, there was criticism against on-line exercise compared to an on-site exercise. Actually, such a feedback was predictable because communication during an on-line exercise is more challenging than communication on-site. Following quotations from the answers illustrate the feedback:

> "It is a beautiful experience, very informative and fun.", "Very interesting and great fun.", "Keep up the good work! Waiting for the Flagship 2.", "Thanks for all attempts and environment. Exercise needs to be held physically, virtual experience is missing a lot!"

Overall, it can be summarised that the implementation of cyber range technical federation was successful. Of course, there were some critics, but it was predictable because people are used to participating in on-site exercises, and an on-line exercise has its own challenges with communication processes and maintaining cyber security situational awareness inside the Blue Team. It is noticeable that this was the first version of the open-source solution and the first time to execute a pan-European on-line exercise with remote participants. In that sense, the results of the showcase were positive and form a great foundation for the Flagship 2 exercise conducted during the next year.

5 Conclusion

During this case study the implementation of cyber range technical federation was tested during the Flagship 1 on-line cyber security exercise. The technical implementation was based on state-of-the art open-source SD-WAN technology with great success, as it enabled the parties to join the on-line exercise. SD-WAN was chosen based on the features of security and flexible deployment options compared to, e.g., site-to-site IPSec or VPN solutions which are commonly used for federation purposes. It is noticeable that the on-line cyber security exercise is technically much more demanding than the on-site cyber security exercise because the learning audience of the exercise is located remotely across the globe. This raises requirements for the usability of the technical environment in order to create a satisfactory user experience. The on-line cyber security exercise has special challenges also with exercise control functionalities, communication inside the Blue Team, communication between the Blue Teams and maintaining situational awareness.

The Flagship 1 exercise demonstrated a good performance and capability of chosen state-of-the-art technologies and the implementation of cyber range technical federation was a success. A participant survey was conducted and analysed as a reliability assessment of the usefulness of the showcased environment. Closed questions of the participant survey were analysed as quantitative data while open questions of the participant survey were analysed by qualitative methods. As a summary, the analysis indicated that the participants were pleased with the exercise experience, and the technical implementation obtained positive feedback. In conclusion, the first version of the cyber range technical federation forms a good foundation for the next phase of the implementation during the Flagship 2 exercise in January 2022.

Acknowledgements This research is funded by *Cyber Security Network of Competence Centres for Europe (CyberSec4Europe)* -project of the Horizon 2020 SU-ICT-03-2018 program. The authors would like to thank Ms. Tuula Kotikoski for proofreading the manuscript.

References

1. Amazon Web Services, Inc. Amazon Web Services (AWS). https://aws.amazon.com/. Accessed 8 Apr 2021
2. Chen Z, Yan L, He Y, Bai D, Liu X, Li L (2018) Reflections on the construction of cyber security range in power information system. In: 2018 IEEE 3rd advanced information technology, electronic and automation control conference (IAEAC), pp 2093–2097. https://doi.org/10.1109/IAEAC.2018.8577685
3. Cyber Security Network of Competence Centres for Europe-project: Cyber Security for Europe (CS4E). https://cybersec4europe.eu/. Accessed 7 Apr 2021
4. Debatty T, Mees W (2019) Building a cyber range for training cyberdefense situation awareness. In: 2019 international conference on military communications and information systems (ICMCIS), pp 1–6. https://doi.org/10.1109/ICMCIS.2019.8842802
5. Deckard GM (2018) Cybertropolis: breaking the paradigm of cyber-ranges and testbeds. In: 2018 IEEE international symposium on technologies for homeland security (HST), pp 1–4. https://doi.org/10.1109/THS.2018.8574134
6. Di Tizio G, Massacci F, Allodi L, Dashevskyi S, Mirkovic J (2020) An experimental approach for estimating cyber risk: a proposal building upon cyber ranges and capture the flags. In: 2020 IEEE European symposium on security and privacy workshops (EuroS PW), pp 56–65. https://doi.org/10.1109/EuroSPW51379.2020.00016
7. Drisko J, Maschi T (2016) Content analysis. Oxford University Press, New Yourk, NY
8. Elliott V (2018) Thinking about the coding process in qualitative data analysis. Qual Rep 23:2850–2861
9. European Commission (2019) Four EU pilot projects launched to prepare the European cybersecurity competence network. https://digital-strategy.ec.europa.eu/en/news/four-eu-pilot-projects-launched-prepare-european-cybersecurity-competence-network. Accessed 9 Apr 2021
10. European Commission (2020) Joint communication to the European parliament and the council: the EU's cybersecurity strategy for the digital decade. https://eur-lex.europa.eu/legal-content/EN/ALL/?uri=JOIN:2020:18:FIN
11. European Commission (2020) Proposal for a directive of the European parliament and of the council on the resilience of critical entities. https://eur-lex.europa.eu/legal-content/EN/TXT/?uri=COM:2020:829:FIN
12. European Cyber Security Organisation (ECSO) (2020) Understanding cyber ranges: from hype to reality. https://www.ecs-org.eu/documents/uploads/understanding-cyber-ranges-from-hype-to-reality.pdf
13. European Defence Agency, EDA EDA's pooling & sharing-factsheet. https://eda.europa.eu/docs/default-source/eda-factsheets/final-p-s_30012013_factsheet_cs5_gris. Accessed 9 Apr 2021
14. European Defence Agency, EDA (2017) Cyber ranges: EDA's first ever cyber defence pooling & sharing project launched by 11 member states. https://www.eda.europa.eu/info-hub/press-centre/latest-news/2017/05/12/cyber-ranges-eda-s-first-ever-cyber-defence-pooling-sharing-project-launched-by-11-member-states. Accessed 7 Apr 2021
15. European Defence Agency, EDA (2018) Cyber ranges federation project reaches new milestone. https://www.eda.europa.eu/info-hub/press-centre/latest-news/2018/09/13/cyber-ranges-federation-project-reaches-new-milestone. Accessed 7 Apr 2021
16. European Defence Agency, EDA (2019) EDA cyber ranges federation project showcased at demo exercise in Finland. https://www.eda.europa.eu/info-hub/press-centre/latest-news/2019/11/07/eda-cyber-ranges-federation-project-showcased-at-demo-exercise-in-finland. Accessed 7 Apr 2021
17. Ferguson B, Tall A, Olsen D (2014) National cyber range overview. In: 2014 IEEE military communications conference, pp 123–128. https://doi.org/10.1109/MILCOM.2014.27

18. Graziano A (2020) About federation of cyber ranges, market places and technology innovation. https://www.linkedin.com/pulse/federation-cyber-ranges-market-places-technology-almerindo-graziano/. Accessed 9 Apr 2021
19. He Y, Yan L, Liu J, Bai D, Chen Z, Yu X, Gao D, Zhu J (2019) Design of information system cyber security range test system for power industry. In: 2019 IEEE innovative smart grid technologies—Asia (ISGT Asia), pp 1024–1028. https://doi.org/10.1109/ISGT-Asia.2019.8881739
20. Hsieh HF, Shannon SE (2005) Three approaches to qualitative content analysis. Qual Health Res 15(9):1277–1288. https://doi.org/10.1177/1049732305276687
21. JAMK University of Applied Sciences, Institute of Information Technology (2021) Coming soon—a cybersecurity exercise that emphasizes learning and cooperation. https://jyvsectec.fi/2021/01/cybersec4europe-projects-cybersecurity-exercise-on-january/. Accessed 7 Apr 2021
22. JAMK University of Applied Sciences, Institute of Information Technology/JYVSECTEC RGCE cyber arena. https://jyvsectec.fi/rgce. Accessed 8 Apr 2021
23. Karjalainen M, Kokkonen T (2020) Comprehensive cyber arena; the next generation cyber range. In: 2020 IEEE European symposium on security and privacy workshops (EuroSi&PW), pp 11–16. https://doi.org/10.1109/EuroSPW51379.2020.00011
24. Karjalainen M, Kokkonen T, Taari N (2022) Key elements of on-line cyber security exercise and survey of learning during the on-line cyber security exercise. Springer International Publishing, Cham, pp 43–57. https://doi.org/10.1007/978-3-030-91293-2_2
25. National Institute of Standards and Technology NIST Cyber ranges. https://www.nist.gov/system/files/documents/2018/02/13/cyber_ranges.pdf. Accessed 13 Jan 2020
26. Nevavuori P, Kokkonen T (2019) Requirements for training and evaluation dataset of network and host intrusion detection system. In: Rocha Á, Adeli H, Reis LP, Costanzo S (eds) New knowledge in information systems and technologies. Springer International Publishing, Cham, pp 534–546
27. North Atlantic Treaty Organization, NATO (2017) Smart defence. https://www.nato.int/cps/en/natolive/topics_84268.htm. Accessed 9 Apr 2021
28. Piispanen J (2018) Technical specification for federation of cyber ranges. Master's thesis, JAMK University of Applied Sciences. http://urn.fi/URN:NBN:fi:amk-2018121722010
29. Piispanen J, Päijänen J (2021) Evaluation report on integration demonstration. https://cybersec4europe.eu/wp-content/uploads/2021/08/D7.3-Evaluation-report-on-integration-demonstration-v1.3_submitted.pdf
30. Päijänen J, Viinikanoja J, Piispanen J (2021) Flagship 1. https://cybersec4europe.eu/wp-content/uploads/2021/06/D6.4-Flagship-1-v1.1-submitted.pdf
31. Secretariat of the Security Committee (2019) Finland's cyber security strategy, Government Resolution 3.10.2019. https://turvallisuuskomitea.fi/wp-content/uploads/2019/10/Kyberturvallisuusstrategia_A4_ENG_WEB_031019.pdf
32. Shangting M, Quan P (2021) Industrial cyber range based on QEMU-IOL. In: 2021 IEEE international conference on power electronics, computer applications (ICPECA), pp 671–674. https://doi.org/10.1109/ICPECA51329.2021.9362692
33. Shepherd LA, de Paoli S, Conacher J (2020) Human-computer interaction considerations when developing cyber ranges. Int J Inf Secur Cybercrime 9(2):28–32. https://doi.org/10.19107/IJISC.2020.02.04
34. Suni E, Piispanen J, Nevala J, Päijänen J, Saharinen K (2020) Report on existing cyber ranges, requirements. https://cybersec4europe.eu/wp-content/uploads/2020/09/D7.1-Report-on-existing-cyber-ranges-and-requirement-specification-for-federated-cyber-ranges-v1.0_submitted.pdf
35. The White House, signed by President Donald J. Trump (2018) National cyber strategy of the United States of America. https://trumpwhitehouse.archives.gov/wp-content/uploads/2018/09/National-Cyber-Strategy.pdf
36. Tian Z, Cui Y, An L, Su S, Yin X, Yin L, Cui X (2018) A real-time correlation of host-level events in cyber range service for smart campus. IEEE Access 6:35355–35364. https://doi.org/10.1109/ACCESS.2018.2846590

37. Ukwandu E, Farah MAB, Hindy H, Brosset D, Kavallieros D, Atkinson R, Tachtatzis C, Bures M, Andonovic I, Bellekens X (2020) A review of cyber-ranges and test-beds: current and future trends. Sensors 20(24). https://doi.org/10.3390/s20247148
38. Urias VE, Stout WMS, Van Leeuwen B, Lin H (2018) Cyber range infrastructure limitations and needs of tomorrow: a position paper. In: 2018 international Carnahan conference on security technology (ICCST), pp 1–5. https://doi.org/10.1109/CCST.2018.8585460
39. ZeroTier Inc. ZeroTier global—area networking. https://www.zerotier.com/. Accessed 29 Apr 2021
40. Zhang Z, Lu G, Zhang C, Gao Y, Wu Y, Zhong G (2020) Cyfrs: a fast recoverable system for cyber range based on real network environment. In: 2020 information communication technologies conference (ICTC), pp 153–157. https://doi.org/10.1109/ICTC49638.2020.9123273

Collaborative Private Classifiers Construction

Mina Alishahi and Vahideh Moghtadaiee

Abstract Cyber-physical systems (CPS) are smart computer systems that control or monitor machines through computer-based algorithms, which are vulnerable to both cyber and physical threats. Similar to the growing number of applications, CPS also employ classification algorithms as a tool for data analysis and continuous monitoring of the system. While the utility of data is significantly important in building an accurate and efficient classifier, a free access to original (raw) format of data is a crucial challenge due to privacy constraints. Therefore, it is tremendously important to train classifiers in a private setting in which the privacy of individuals is protected, while data remains still practically useful for building the model. In this chapter, we investigate the application of three privacy preserving models, namely anonymization, Differential Privacy (DP), and cryptography, to privatize data and evaluate the performance of two popular classifiers, Naïve Bayes and Support Vector Machine (SVM) over the protected data. Their performances are compared in terms of accuracy, training construction costs on the same data and in the same private environment. Finally, comprehensive findings on constructing the privacy preserved classifiers are outlined. The attack models against the training data and against the private classifier models are also discussed.

M. Alishahi (✉)
Department of Computer Science, Open Universiteit, Heerlen, The Netherlands
e-mail: mina.sheikhalishahi@ou.nl

V. Moghtadaiee
Cyberspace Research Institute, Shahid Beheshti University, Tehran, Iran
e-mail: v_moghtadaiee@sbu.ac.ir

T. Dimitrakos et al. (eds.), *Collaborative Approaches for Cyber Security in Cyber-Physical Systems*, Advanced Sciences and Technologies for Security Applications,
https://doi.org/10.1007/978-3-031-16088-2_2

1 Introduction

Cyber-physical systems (CPS) can be defined as systems which incorporate several components such as networks, sensors, computers, and digital monitoring devices into the physical infrastructure aiming to control and monitor the operation of this infrastructure autonomously. In modern real-world scenarios, CPS can be extensively used in smart grids, medical monitoring systems, robotics, autonomous vehicles, smart cities, military, soil treatment plants, smart homes, and water treatment plants [49]. As the implementation of CPS includes both cyber and physical components, these systems can be the target of both cyber and physical security threats.

To prevent such kind of security threats, *classification* algorithms are increasingly used within CPS to enable a continuous monitoring of the system and to enhance quality of service, privacy, and security measures [49, 50]. Classification is the task of identifying the class label of a new instance, based on a formerly obtained model on training set of instances with known class labels. Under the condition that classification algorithms are trained on a large amount of reliable and meticulous data, they can be used to infer additional knowledge and intelligence about the system and raise an on-time alarm in the case of security threats.

This assumption requires the free access of an analyzer to data which might not be available in a central location. In many real-world scenarios the training data is generated and governed by different entities who are unwilling to share their data with the data analyzer as it might contain privacy-sensitive information, and its disclosure might raise the data owners' privacy concerns [52]. Consider, for instance, a situation in which a hospital owns a dataset describing patients' information, including age, address, gender, symptoms, and diseases. A classifier trained on this dataset might leak sensitive information about any individual among the patients. Moreover, sharing sensitive information might potentially cause security threats, e.g., revealing when a person is not at home to a burglar [3]; financially harm a person, e.g., when the insurance companies have access to the patients' medical records [45]; reveal the habits and interests of a person, e.g., when the information of the locations a person has visited is accessible [16]; and negatively affect a company's reputation, e.g., when an adversary has access to the customers' personal information [10].

To assure the protection of individuals' confidential information, several countries have enacted data protection and privacy regulations that impose privacy restrictions on the process of collecting and analyzing the data. To name a few, (e.g., GDPR,[1] PCI DSS,[2] HIPAA[3]) impose strict privacy requirements regulating the collection, processing and using personal data that prevent harm on personality rights.

To address the privacy problem, some studies propose to protect data privacy such that the desired data analysis can still be performed over protected data [8, 19]. In particular, the privacy preserving construction of a classifier over distributed data has obtained considerable attention [8, 18, 22, 54, 67]. In general, the proposed

[1] https://gdpr-info.eu.

[2] https://www.pcisecuritystandards.org.

[3] https://www.hhs.gov/hipaa.

solutions employ different privacy preserving approaches to securely build classifiers on decentralized data. The main privacy preserving solutions can be based on three techniques: the data anonymization, Differential Privacy (DP), and cryptography.

Anonymization techniques basically replace the values in the data with a more general representation before the dataset is published, e.g., k-anonymity [47, 56], ℓ-diversity [37], and t-closeness [30]. These techniques have been employed to generate anonymized labeled dataset so that it could be published for the construction of the classifiers [24]. One application of anonymization techniques in data classification is based on the two-party Setting, in which one entity (data provider) owns the data and another entity (data analyzer) is interested in training a classifier on this data. The data analyzer needs to know which classifier outperforms the other classifiers on the shared anonymized data. This knowledge would enable the analyzer to decide which classifier should be trained based on the anonymization technique employed, the dataset properties, and the desired performance metric. The anonymization techniques have been criticized for not being rigorous enough in protecting the individuals' confidential information.

Differential Privacy (DP) is considered as a significant privacy standard for data privatization [14]. It addresses the weaknesses of anonymization techniques by limiting the revelation of private information of individual records. The strict privacy guarantee given by DP has led to its broad application in the field of privacy preserving data analysis. One of the main scenarios in which DP is applied is to introduce noise during the training of classifiers, where the noise is scaled according to the sensitivity of the training algorithm.

In this setting, multiple classification algorithms have been extended to incorporate differential privacy with private classifiers' learning, e.g., Nearest Neighbor [22], Naïve Bayes [59], Support Vector Machine (SVM) [11], and Decision Tree [25].

Encryption or cryptography-based methods have also been employed to securely train a classifier over protected data [29]. For instance, Homomorphic Encryption (HE) has been well adopted to learn classifiers over encrypted data, e.g., the logistic regression algorithm [6] and Random Forest classifiers [28] have been privately trained over distributed data. However, the cryptographic-based methods are not scalable both in terms of execution runtime and bandwidth usage [43] and they are designed only for a specific classification algorithm [28].

While each one of these privacy preserving models has served as effective solution for training a classifier on private data, there is still a lack of understanding of the impact of different inputs in training private classifiers. This knowledge enables the data provider and data analyzer to decide which classifier be incorporated to which privacy preserving model to be privately trained considering (i) dataset properties, such as dataset size, (ii) structural properties of the classification algorithm, (iii) the amount of privacy level required, (iv) the accuracy of different classification algorithms when trained over private data, and (v) the impact of changes in privacy parameters in different settings on the classifiers performance, by taking into account (vi) the resource limitations (bandwidth usage) and (vii) time constraints of the application domain.

To this end, this chapter provides a comparison of the impact of data privacy in the context of constructing a classifier over anonymized data, differentially private data, and encrypted data. For data anonymization, applying k-anonymity, ℓ-diversity and t-closeness techniques are investigated. For data perturbation, an ε-DP setting is analyzed, where noise is added to the analyst's queries during the training of the classifier. For cryptographic techniques, Secure Function Evaluation particularly the ABY library is explored. In addition, we indicate how Naïve Bayes and SVM classifiers can be constructed over private inputs and we compare their performance in terms of accuracy, training construction costs on the same dataset and in the same private environment. Finally, important findings are outlined on how the privacy techniques can be used to construct classifiers in a private setting. There is also a discussion regarding the main attack models that target the training dataset and private classifier models.

2 Privacy Models

This section provides preliminary concepts used in this study including the basics of three major privacy preserving techniques, i.e., anonymization, DP, and encryption-based techniques.

2.1 Anonymization

The aim of anonymizing data is to ensure that personal information cannot be unambiguously linked back to a particular individual. For this study, we consider three popular anonymization techniques: k-anonymity [47, 56], ℓ-diversity [37], and t-closeness [30].

We consider a dataset comprising a set of records, where each record corresponds to one individual. Each record is described by a number of attributes, which can be divided into three categories: (1) *identifiers* that uniquely identify the individuals, e.g., social security number, (2) *quasi-identifier* (QI) attributes in which taking all values together can be used to potentially identify an individual, e.g., zip-code, birth-date, and gender, (3) *sensitive attributes* (SA) which contains very personal and sensitive information that an adversary is not allowed to discover the values of that attribute for any individual, e.g., a patient's disease or an employer's salary. A group of records in an anonymized dataset is called *equivalence class* when they have identical values in QI attributes.

2.1.1 k-Anonymity

A release of data is said to satisfy k-anonymity if each record in the release cannot be distinguished from at least $k-1$ other records in the release with respect to QIs [56]. k-anonymity is susceptible to some attacks, e.g. *homogeneity* and *background knowledge* attacks.

$$priv_{KA} \equiv k, \ where\, \forall E : |E| \geq k \tag{1}$$

where E is the equivalence class, and $priv_{KA} \equiv k$ denotes that the associated dataset respects k-anonymity. Consider, for instance, the small dataset on the original raw data of several disease diagnoses shown in Fig. 1a. To protect the privacy of the patients, the identifiers, e.g. patients' name, is already deleted from the raw data. The QIs, *zip-code* and *age*, are anonymized so that the dataset satisfies 2-anonymity (Fig. 1b).

2.1.2 ℓ-Diversity

The ℓ-diversity model addresses some of the weaknesses of k-anonymity. In particular, k-anonymity does not protect the values of SAs, specifically when the values in an equivalence class are identical. To address this drawback, the ℓ-diversity model introduces constraints on intra-group diversity for SAs [37]. Formally,

$$priv_{LE} \equiv \ell, \ where\, \forall E : H(S_E) \geq \log(\ell) \tag{2}$$

where S_E is the SAs appeared in the equivalence class E and $H(S_E)$ denotes the entropy of these values. Figure 1c reports an anonymized version of the dataset in Fig. 1a satisfying 3-diversity, where both *Salary* and *Disease* is considered SAs. We can observe that three distinct sensitive values appears in every anonymous equivalence class. However, ℓ-diversity does not consider the semantic closeness of the distinct values in a SA. This problem is addressed by t-closeness.

2.1.3 t-Closeness

An equivalence class is said to satisfy t-closeness if the distance between the distribution of a SA in this group and the distribution of the attribute in the whole dataset is not greater than a given threshold t. Accordingly, the dataset satisfies t-closeness if all equivalence classes in the dataset satisfy t-closeness [30]. Formally,

$$priv_{TC} \equiv t, \ where\, \forall E : d(S, S_E) \leq t \tag{3}$$

where S_E and S show the distribution of sensitive values in equivalence class E and in the whole dataset, respectively. Figure 1d shows a dataset satisfying 0.167-closeness w.r.t. *Salary* and 0.278-closeness w.r.t. *Disease*.

	ZIP Code	Age	Salary	Disease
1	47677	29	3K	gastric ulcer
2	47602	22	4K	gastritis
3	47678	27	5K	stomach cancer
4	47905	43	6K	gastritis
5	47909	52	11K	flu
6	47906	47	8K	bronchitis
7	47605	30	7K	bronchitis
8	47673	36	9K	pneumonia
9	47607	32	10K	stomach cancer

(a)

	ZIP Code	Age	Salary	Disease
1	47***	*	[3K- 4K]	gastric ulcer
2	47***	*	[3K- 4K]	gastritis
3	47***	*	[5K - 6K]	stomach cancer
4	47***	*	[5K - 6K]	gastritis
6	47***	*	[7K - 8K]	bronchitis
7	47***	*	[7K - 8K]	bronchitis
5	47***	*	[9K - 11K]	flu
8	47***	*	[9K - 11K]	pneumonia
9	47***	*	[9K - 11K]	stomach cancer

(b)

	ZIP Code	Age	Salary	Disease
1	476**	2*	3K	gastric ulcer
2	476**	2*	4K	gastritis
3	476**	2*	5K	stomach cancer
4	4790*	≥ 40	6K	gastritis
5	4790*	≥ 40	11K	flu
6	4790*	≥ 40	8K	bronchitis
7	476**	3*	7K	bronchitis
8	476**	3*	9K	pneumonia
9	476**	3*	10K	stomach cancer

(c)

	ZIP Code	Age	Salary	Disease
1	4767*	< 40	3K	gastric ulcer
3	4767*	< 40	5K	stomach cancer
8	4767*	< 40	9K	pneumonia
4	4790*	≥ 40	6K	gastritis
5	4790*	≥ 40	11K	flu
6	4790*	≥ 40	8K	bronchitis
2	4760*	< 40	4K	gastritis
7	4760*	<40	7K	bronchitis
9	4760*	< 40	10K	stomach cancer

(d)

Fig. 1 Examples of k-anonymity, ℓ-diversity, and t-closeness, **a** the original dataset, **b** 2-anonymity, **c** 3-diversity **d** 0.167-closeness for *Salary* and 0.278-closeness for *Disease*

2.2 Differential Privacy

By inserting an appropriate amount of noise, DP keeps real-time or statistical data private while keeping an acceptable balance between accuracy and privacy [20]. In DP, if two datasets differ in at most one row, they are called *adjacent*. Let privacy budget, $\varepsilon \in \mathbb{R}_{\geq 0}$, and f be an algorithm operating on datasets. We say f satisfies ε-Differential Privacy if for adjacent datasets D, D' and all sets of possible outputs M, we have:

$$\mathbb{P}(f(D) \in M) \leq e^{\varepsilon}\mathbb{P}(f(D') \in M). \tag{4}$$

By ensuring that the probability distributions on the output space originating from two input datasets cannot differ too much, ε-DP provides plausible deniability about any row's true value, even if all other rows are compromised. The lower ε, the stronger privacy ε-DP guarantees. To ensure privacy in the classifier learning setting, we demand that a classifier training algorithm satisfies ε-DP.

2.3 Encryption

Secure function evaluation (*SFE*) is a family of cryptographic constructions that enable several parties to compute a function on their private inputs without revealing any information except the result of the function. In this work, we implement SFE in two-party setting using the ABY framework [13], which is an efficient framework defined based on the secure two-party computation that allows pre-computing cryptographic operations and provides the conversions between various secure computation schemes based on pre-computed oblivious transfer extensions. In particular, ABY provides the constructions for Arithmetic sharing, Boolean sharing, and Yao's garbled circuits. For the definition of our protocols, we only use Arithmetic sharing, since they provide efficient constructions for our classifiers.

Given two parties, *server* and *client*, and integer inputs a and b, ABY first creates secret shares for each party, and then securely computes f on the secret shares. Secret shares of each party is represented as $\langle a\rangle_1$, $\langle a\rangle_2$ and $\langle b\rangle_1$, $\langle b\rangle_2$ such that $\langle a\rangle_1 + \langle a\rangle_2 \equiv a \mod 2^\varsigma$ and $\langle b\rangle_1 + \langle b\rangle_2 \equiv b \mod 2^\varsigma$. To design our secure computation protocols, we adopt two Arithmetic gates from the ABY framework as building blocks, namely the *Addition* and *Multiplication* gates.

Addition Gate: The addition gate overloads integer addition such that the result is equal to the addition of two secret shared inputs in modulus 2, where ς is the bit size of the inputs. Given two secret shared inputs $\langle a\rangle$ and $\langle b\rangle$, the addition gate is represented as:

$$\langle a + b\rangle = \langle a\rangle + \langle b\rangle \mod 2^\varsigma. \tag{5}$$

Multiplication Gate: The multiplication gate overloads integer multiplication such that the result is equal to the multiplication of two secret shared inputs in modulus 2^ς. Given two secret shared inputs $\langle a\rangle$ and $\langle b\rangle$, the multiplication gate is represented as:

$$\langle a \times b\rangle = \langle a\rangle \times \langle b\rangle \mod 2^\varsigma. \tag{6}$$

where the multiplication gate is defined based on additive Homomorphic encryption over secret shares of data [13].

It should be noted that while here we only present the two-party secure functional evaluation scheme as a cryptographic technique, our findings about the applications of this scheme in training private classifiers can be generalized to a broader set of cryptographic schemes, e.g., Paillier cryptosystem.

3 Literature Review

Privacy preserving data classification can be categorized into three main groups: (i) approaches aiming to generate anonymized datasets for training classifiers. (ii) iterative protocols for the construction of a classifier based on data perturbation techniques, and (iii) secure computation protocols for the construction of a classifier over protected data. The most related work on these three approaches are discussed below.

3.1 On Anonymization

Data anonymization has become a widely investigated research direction in an effort to protect individuals' privacy when data is supposed to be released publicly. k-anonymity, which was proposed as the initial definition of anonymity [47, 56], has been extended to new additional constraints such as ℓ-diversity [37] and t-closeness [30]. The proposed approaches have been optimized in terms of a generic measurement with no emphasis on the utility of anonymized data for classification. For instance, in [32], a novel anonymization technique named *slicing* is proposed, which handles high-dimensional data and improves the data structure compared to preliminary generalization technique. Authors in [44] also suggest a hybrid generalization technique which provides a trade-off between utility and privacy by data relocation.

Moreover, design and application of anonymization techniques when data utility is critical for data classification has been investigated in several studies. A new anonymization approach based on rough set theory to measure data quality for accurate classifier construction is introduced in [63]; it guides the anonymization process through combining rough set theory and attribute value taxonomies. Authors in [40, 51] obtain a trade-off between privacy and data utility by employing an appropriate feature suppression for publishing the private dataset to be further used for classification. The focus of these approaches is to create anonymized datasets such that the features successfully distinguish the class labels. The performance of a specific classifier over the anonymized datasets, on the other hand, has not been sufficiently investigated. Authors in [18] try to embed (k-anonymity) within the decision tree induction process, which provides higher accuracy than the case when the data is anonymized first and used for inducing the tree afterwards. A similar methodology has been proposed in [12] for embedding anonymization within the association rule mining algorithm. The researchers in [55] also compare different k-anonymization algorithms and investigate their effects on machine learning results. The effect of anonymization on the classifiers' accuracy is explored in [38, 39] by conducting a series of experiments and applying four different classifiers to the Adult dataset.

Beside the aforementioned studies, some researchers assess the effect of dataset features on data anonymization. In [41], for instance, novel methods are proposed to identify which features of documents are necessary to be changed to accomplish the document anonymization. Authors in [9] also investigate whether generalization and suppression of Quasi-Identifier (QI) features offer any benefit over trivial sanitization which simply separates QI features from Sensitive Attributes (SA) ones.

3.2 *On Differential Privacy*

Another line of research has proposed the application of data perturbation techniques for the construction of privacy preserved classifiers [31, 46] to provide a model that does not leak undesirable information about the training data. In most cases, DP is used for the purpose of data perturbation to build a specific classification algorithm [21]. In order to guarantee DP in DP-based classification, an entity (quester) wants to build a classifier using non-public data of other entity (data provider), where she answer the quester's queries by adding noise to the dataset [17]. DP also can be applied to the table of data before being released for constructing a classifier [4].

DP technique has been used for the construction of Naïve Bayes classifiers [31] and SVM classifiers [46], where a single provider has centralized access to a dataset and would like to release a classifier while protecting privacy. For this purpose, the dataset queries in the standard classifier algorithm is replaced with differentially private ones. The authors in [66] try to enhance the results by using a DP technique that lowers the amount of random noise on each query, while keeping the same level of privacy. Other works show how Naïve Bayes classifiers can be trained over noisy data [2], or with data transformation and hiding using Randomized Response [58]. The authors in [26] propose a method, namely (ε, δ)-DP, which is a weaker form of privacy but can be applied to a wider range of kernel functions. Differentially private Decision Tree algorithms is also reviewed in [17]. For instance, the differentially private data is used to train Decision Tree classifier in [7] instead of using a non-private data.

In addition to DP, a *Local Differential Privacy* (LDP)-based approach is also employed for classification algorithm in data centers [15]. In LDP, each entity (as one data owner) locally adds a controlled noise to her data before sending it to the data centers for aggregation and creation of a dataset, since the data aggregator is considered not trusted. LDP provides strong privacy in training the classifiers [21], while the utility cost might be higher especially when the number of all data owners is not high enough. In [64], LDP has been employed with a non-interactive obfuscated dataset to train Naïve Bayes classifier.

Noise in DP classification is not always added to the data, it can be also added to the parameters of the trained classifier with non-private data before the classifier's model is publicly released. This approach, however, usually shows lower accuracy. For instance, the authors of [11] perturb the optimal hyperplane of the SVM classifier. Furthermore, in a Random Forest classifier, the noise has been added to the leaves' class predictions taken from a database [25]. However, multiple trees is needed to get acceptable accuracy even though this method requires fewer queries per tree.

In order to better realize the impact of different ε values and various relaxations of DP on both utility and privacy, the performance evaluation of DP on two machine learning algorithms is also discussed in [27].

3.3 On Encryption

Data encryption is also applied for privacy preserving classification machine learning. They mainly focus on designing a framework for the secure construction of a specific classifier over distributed data using cryptographic techniques [28, 33]. A cryptographic tool called additive homomorphic proxy aggregation scheme is used in [35] to train Naïve Bayes classifier over distributed data. Besides, a blockchain-based privacy preserving SVM classification between mutually distrustful data owners over horizontally distributed mobile crowd sensed data is designed in [33]. Authors in [65] also try to train the SVM classifier among more than two parties, over vertically distributed data, using the secure sum protocol. The homomorphic encryption is applied in training logistic regression algorithm, where sensitive information is protected both during the training and prediction process [6]. A homomorphic k-Nearest Neighbours determination algorithm is also introduced in [68] to be further employed in machine learning classifiers. Moreover, the ID3 decision tree algorithm is privately trained over vertically [57] and horizontally distributed data [60] when data is distributed between two or more parties. Researchers in [28] train a bagging classification, e.g., Random Forest, over horizontally and vertically partitioned data. A secure prediction system that allows two parties to execute neural network on their combined data is also proposed in [42].

Despite all the the research have been done, an analysis and comparison of the constructed private classifier's accuracy, complexity and computation costs under different conditions is still missing. This chapter provides a comparison on private classifiers' construction using the aforementioned privacy preserving models.

4 Problem Statement

An explosive growth in data availability arising from the development of 5G/6G technologies, combined with the ever-increasing access of everyone to Internet, promotes the application of classification algorithms in many domains. These applications include, but are not limited to, the data analysis in CPS, health-care, intelligent transportation and logistics, smart grid, smart cities, smart home, localization and tracking, financial, and smart marketing [1, 2, 16, 48, 62]. To complete the classification task successfully, we need to protect the individuals' privacy. This is because the leakage of participants' information might reveal their sensitive information, e.g., their daily activities, religious/political points of view, and social habits [39].

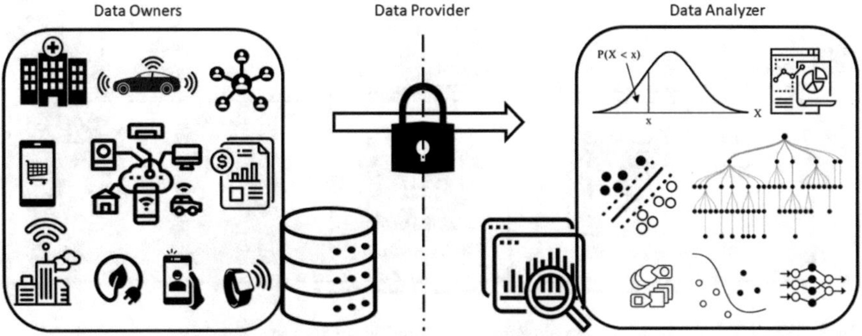

Fig. 2 The reference architecture of private classifier construction

In this study, we investigate the application of different privacy preserving techniques in constructing the classification models over privacy protected data. Figure 2 depicts a high level representation of our setting. On one side, different kinds of data are generated and collected in original (raw) format, and on the other side, the classifier is trained without accessing to the original data as privacy protected methods are already applied on them before getting to the classifiers. This architecture is constituted of three main entities:

- *Data Owner* is an individual who generates the data and is the main owner of data whose privacy matters, e.g., a person health-related sensor data, online shopping records, visited locations information, and energy consumption pattern.
- *Data Provider* is an entity who collects the data owners' (raw or noisy) data, e.g., a hospital, an e-commerce company, and a bank.
- *Data Analyzer* is an entity that neither the data owners nor the data provider can fully trust, but it has knowledge and expertise in constructing a classifier out of collected data.

Depending on the privacy preserving models used for private classifiers construction, the roles of these three entities might change (Sects. 4.1, 4.2 and 4.3). Given that, the input data properties, the selection of the privacy preserving model (along with the associated parameters), and the structure of classification algorithm affect the private classifiers' accuracy, the building runtime and bandwidth usage, and the model security. We investigate and compare private classifiers constructed over protected data in different settings.

4.1 Classifiers Over Anonymized Data

We assume a data provider wants to release a dataset D to a data analyzer for modeling a classifier on this data. The data provider wants to protect the dataset against linking an individual to sensitive information using an anonymization approach. The data

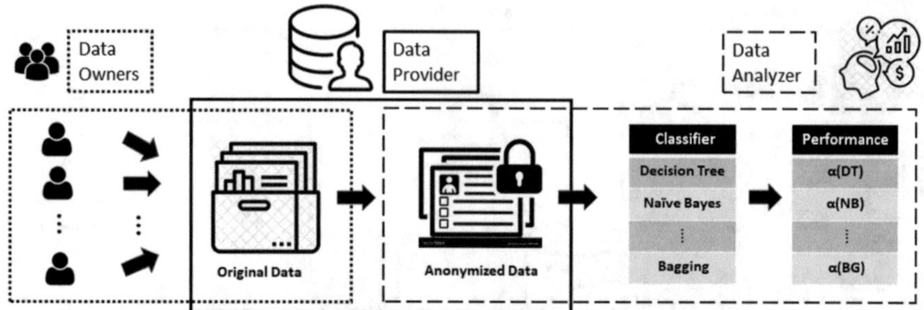

Fig. 3 The architecture of private classifier construction on anonymized data

analyzer who has access to the anonymized version of data is interested in training a classifier. However, the data analyzer has no knowledge which classifier should be selected on the published anonymized data. An overview of this communication model with these two entities is presented in Fig. 3. The figure shows that data is anonymized before being sent for training the classifier. In this setting, the data owners fully trust the data provider, where in real world scenario this data provider can be considered an organization who collects its costumers/clients/employees data [5].

4.2 Classifiers Over Differentially Private Data

Given a privacy level ε and a dataset D, we need to determine the ε-DP classifier training algorithm Q which maximizes the accuracy of the classifier $Q(D)$. Classifiers can be trained by retrieving information from a dataset through numerical queries. In this case, ε-DP is ensured when the responses to queries satisfy DP. On a single query, DP can be incorporated as follows. Let φ be a numerical function on datasets, and let $\bar{s} := \max |\varphi(D) - \varphi(D')|$, where the maximum is taken over all adjacent datasets D, D'. Suppose a query asks for $\varphi(D)$. Then the response is ε-DP if

$$L(\varphi, \varepsilon) = \varphi(D) + \text{Lap}(0, \bar{s}/\varepsilon), \tag{7}$$

where $\text{Lap}(0, \bar{s}/\varepsilon)$ is a zero mean Laplace random variable with the scaling parameter $\bar{s}/\varepsilon$. Generally, we need positive responses, in which case we will use $L^+(\varphi, \varepsilon) = \max\{L(\varphi, \varepsilon), \alpha\}$, where α is a small positive number that should be substantially smaller than $\varphi(D)$. Since most of our queries are counts and therefore integers, we use $\alpha = 10^{-5}$ throughout the work. The response $L^+(\varphi, \varepsilon)$ is ε-DP as well. These DP responses can be used to construct DP-based classifier. Theoretically, suppose that each row of the dataset is accessed through at most m queries and that the response to each query is $\frac{\varepsilon}{m}$-DP. Then, Q is ε-DP [36].

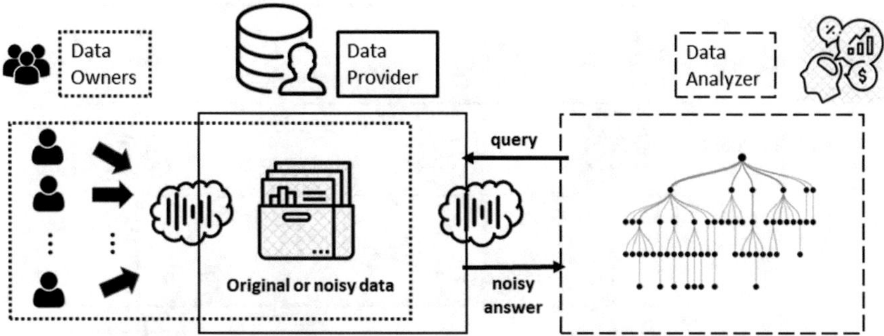

Fig. 4 The architecture of private classifier construction on differentially private data

The DP setting of constructing a classifier in this study is depicted in Fig. 4. As it can be observed, data is owned by data provider who has already collected raw or noisy data from the data owners and it provides noisy answers to the data analyzer's queries.

4.3 Classifiers Over Encrypted Data

In our setting for constructing classifiers over encrypted data, we assume that the data owners trust several data providers. These data providers can be, for instance, several hospitals or e-commerce companies who have access to the users' data. Each data provider P_s $(3 \leq s \leq N)$ holds a dataset D_s. The aim is to learn the classifier from the union of all datasets, i.e. $D = \cup_{s=1}^{N} D_s$. Two entities, named server and client, collaboratively play the role of data analyzer through secure communication. Data providers generate two secret shares of every single value of their data and provide one share to the server and the other one to the client. The shared value is dependent on the function f (classifier) to be implemented. Figure 5 shows the proposed architecture including the main components and their interactions.

We assume a semi-honest adversary model, where all participants are honest and follow the protocols properly with no deviation, but they are curious to learn about other providers' data. This assumption specifically guarantees that the computations do not leak any unintended information. We also assume that the data providers share the correct protected inputs to the corresponding parties at the beginning of the protocol execution.

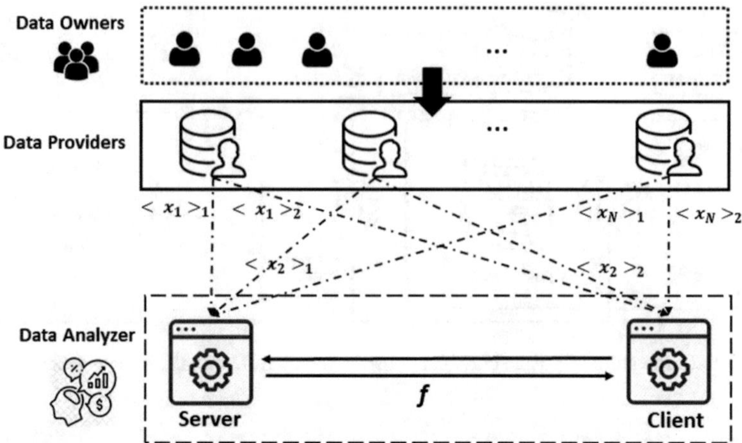

Fig. 5 The architecture of private classifier construction on encrypted data

Table 1 Notations

Notation	Description	Notation	Description
D	Dataset	QI	Quasi-identifier attribute
D_s	Dataset of s'th data provider	SA	Sensitive attribute
P_s	s'th data provider	E	Equivalence class
N	The number of data providers	k	Anonymity value
n	The number of records in D	ℓ	Diversity value
$\mathcal{A}$	The set of attributes	t	Closeness value
x_A	The value of attribute A in record x	ε	Privacy budget
C	Set of class labels	$\bar{x}$	Record x after removing x^c
x^c	Class label of record x	n_{Avc}	Number of rows with $x_A = v$, $x^c = c$

5 Comparing Private Naïve Bayes and SVM Classifiers

A classification algorithm or a classifier is a supervised learning approach, which is used to identify the category of new instances based on a training dataset. Formally, we assume the training dataset D consists of n rows (records) $\mathbf{x}$, where each record is described by a set of attributes $\mathcal{A}$ and a class label from the set of class labels C. More precisely, each row $\mathbf{x}$ is a vector in which each element x_A is a value of attribute $A \in \mathcal{A}$ and a class x^c. We write $\bar{x}$ for the unlabeled row, i.e., for $\mathbf{x}$ with the class x^c removed. The analyst's goal is to create a *classifier*, which can be used to predict the class x'^c of a new unlabeled observation $\bar{x}'$. The notations throughout the paper is listed in Table 1. In this section, we consider two well-known classifiers, named Naïve Bayes classifier and Support Vector Machine, as defined in what follows.

5.1 Naïve Bayes and SVM Classification Algorithms

The basics of Naïve Bayes and SVM Classification Algorithms are presented in this part [1].

5.1.1 Naïve Bayes Classifier

Naïve Bayes classifier is a probabilistic classifier built based on Bayes' theorem and assumes the attributes describing the data are mutually independent. Formally, for an unlabeled data record $\bar{x}$, the conditional probability of $\bar{x}$ being class c is denoted by $p(c|\bar{x})$, and by utilizing Bayes theorem it is calculated as:

$$p(c|\bar{x}) = \frac{p(c)p(\bar{x}|c)}{p(\bar{x})} = \frac{p(c)\prod_{A\in\mathcal{A}} p(x_A|c)}{p(\bar{x})} \tag{8}$$

where $p(c)$ is the probability of class c to occur in the dataset, $p(\bar{x}|c)$ is the conditional probability that $\bar{x}$ occurs given it is labeled c, and $p(\bar{x})$ is the probability of $\bar{x}$ to occur.[4] These probabilities are computed by observing frequencies in the dataset. The classifier assigns a class label $\hat{c}$ to given data with the maximum a posteriori probability as follows:

$$\hat{c} = \arg\max_{c} p(c) \prod_{A\in\mathcal{A}} p(x_A|c) \tag{9}$$

This equation is equivalent to Eq. 8, removing the constant value $p(\bar{x})$ in the denominator. From this formula, it can be inferred that for constructing the Naïve Bayes classifier, it is enough to compute the conditional probabilities and the probability of each class label. For a class c, a categorical attribute A, and a value $v \in A$, the conditional probability $p(x_A = v|c)$ and the probability $p(c)$ are computed as:

$$p(x_A = v|c) = \frac{n_{Avc}}{n_c}, \quad p(c) = \frac{n_c}{n}, \tag{10}$$

where n_{Avc} is the number of rows $\mathbf{x}$ in D with $x_A = v$ and $x^c = c$ and n_c is the number of rows with $x^c = c$. For a numerical attribute A and $z \in \mathbb{R}$, the distribution of x_A given $x^c = c$ is assumed to be normal, and its probability density function is calculated as:

$$p(x_A = z|c) = \frac{1}{\sqrt{2\pi}\sigma_{Ac}} \mathrm{e}^{\frac{(z-\mu_{Ac})^2}{2\sigma_{Ac}^2}}, \tag{11}$$

[4] The second equivalence of (8) is derived by assuming that attributes are independent.

where μ_{Ac} is the mean value of x_A among the rows $\mathbf{x}$ of D with class c, and σ_{Ac} is the standard deviation of these values.

5.1.2 Support Vector Machine (SVM)

algorithm is a classifier defined based on a statistical learning framework, in which the class label is binary, i.e., the classes are ± 1, and the value of each attribute A is numerical. As such, we can represent each $\bar{x}$ as a point in $|\mathcal{A}|$-dimensional space. The aim of the SVM training algorithm is to find a hyperplane in this space that best separates the sets of points corresponding to the two class labels. A hyperplane, represented by a normal vector $\bar{w}$, fails to separate the sets of point at a degree which is measured by:

$$J(\bar{w}, D) = \frac{1}{n} \sum_{i=1}^{n} l_h(x_i^c(\bar{w} \cdot \bar{x}_i)) + \frac{\Lambda}{2} ||\bar{w}||^2, \tag{12}$$

where $x_i^c \in \{\pm 1\}$ is the class of the i-th data record, $\bar{w} \cdot \bar{x}_i$ is the inner product of $\bar{w}$ with the unlabeled data record $\bar{x}_i$, and l_h is the Huber loss function given for a fixed parameter $h > 0$, by

$$l_h(z) = \begin{cases} 0 & \text{if } z > 1 + h, \\ \frac{1}{4h}(1 + h - z)^2 & \text{if } |1 - z| \leq h, \\ 1 - z & \text{if } z < 1 - h. \end{cases} \tag{13}$$

The term $\frac{\Lambda}{2}||\bar{w}||^2$ in (12) is to prevent overfitting. Following [11], we take $h = 0.05$ and $\Lambda = 10^{-2.5}$. The SVM returns the $\bar{w}$ that minimizes (12), i.e., $\hat{w} = \arg\min_{\bar{w}} J(\bar{w}, D)$.

An alternative version of SVM aims to find a hyperplane in an $|\mathcal{A}|$-dimensional space that distinctly classifies the data records. To separate two classes of data records, there are many possible hyperplanes, but the goal is to find the one that has the maximum margin, i.e. the maximum distance between data records of both classes. The $\mathbf{w}$ and b that solve the following optimization problem determine the classifier:

$$\textit{minimize } \|\mathbf{w}\| \quad \text{subject to} \quad x_i^c(\mathbf{w} \cdot \mathbf{x}_i - b) \geq 1 \tag{14}$$

where x_c^i is the class label ($+1$ or -1) of record $\mathbf{x_i}$. Afterwards, to assign a label to a new instance $\mathbf{x}_{new}$, we compute $sgn(\mathbf{w} \cdot \mathbf{x}_{new} - b)$. For the non-linear separable data, a transformation function, denoted by φ, is used to transform the data records to a higher dimensional space where they become linearly separable in the new space. To find the SVM classifier in this case, the following optimization problem is solved:

$$
\begin{aligned}
&\textit{Maximize } f(\gamma_1, \ldots, \gamma_n) \\
&\quad = \sum_{i=1}^{n} \gamma_i - \frac{1}{2} \sum_{i=1}^{n} \gamma_i \sum_{j=1}^{n} x_i^c \gamma_i (\varphi(\bar{x_i}) \cdot \varphi(\bar{x_j})) x_j^c \gamma_j \\
&\quad = \sum_{i=1}^{n} \gamma_i - \frac{1}{2} \sum_{i=1}^{n} \gamma_i \sum_{j=1}^{n} x_i^c \gamma_i k(\bar{x_i}, \bar{x_j}) x_j^c \gamma_j \\
&\text{subject to } \sum_{i=1}^{n} \gamma_i x_i^c = 0, \text{ and } 0 \le \gamma_i \le \frac{1}{2n\lambda} \text{for all } i
\end{aligned}
\tag{15}
$$

where $\gamma_1, \ldots, \gamma_n$ are the regularization parameters of the SVM, λ is the trade-off parameter by increasing margin size and ensuring points are in the correct side, $x_1^c, \ldots, x_n^c$ are the class labels of the data records $\mathbf{x}_1, \ldots, \mathbf{x}_n$, respectively. Also, $k(\bar{x_i}, \bar{x_j})$ is the kernel function between $\bar{x_i}$ and $\bar{x_j}$, where in this work it is defined as the scalar product of two vectors, i.e., $k(\bar{x_i}, \bar{x_j}) = \mathbf{x_i} \cdot \mathbf{x_j}$. After finding the parameters γ_i, we have $\mathbf{w} = \sum_{i=1}^{n} \gamma_i x_i^c \varphi(\mathbf{x_i})$ and $b = \mathbf{w} \cdot \varphi(\mathbf{x_i}) - x_i^c$. To label the new instance $\mathbf{x}_{new}$, it is enough to compute:

$$
sgn(\mathbf{w} \cdot \varphi(\mathbf{x}_{new}) - b) = sgn\big((\sum_{i=1}^{n} \gamma_i x_i^c k(\mathbf{x_i}, \mathbf{x}_{new})) - b\big) \tag{16}
$$

5.2 Naïve Bayes and SVM Classifiers in Private Setting

This section explains how the classifiers need to be changed in the private setting where they are trained over the private data using privacy models. The Naïve Bayes and SVM classifiers over anonymized data, differentially private data, and encrypted data are described bellow in details.

5.2.1 Classifiers Over Anonymized Data

To train classifiers on anonymized datasets, it is assumed that the explicit identifier attributes have been removed from the data, the attribute representing the class label is considered as the SA, and the remaining attributes are considered as QIs [5]. Moreover, the categorical attributes are either removed or replaced with numerical values (e.g. employing one hot encoding, when the conversion is a valid assumption). This is because in the application of anonymization approaches over categorical attributes, the presence of an expert in the field is generally required to provide the taxonomy trees for generalization. The data provider, selects one of the anonymization techniques and the associated value k, ℓ, and t, depending on the required level

Algorithm 1 Construction of ε-DP Naïve Bayes classifier

Require: Privacy budget ε.

Ensure: Prior probabilities $p(c)$; Conditional probabilities $p(x_A = v|c)$ for each class label, categorical attribute, and value of that attribute; Obfuscated mean $\tilde{\mu}_{Ac}$ and standard deviation $\tilde{\sigma}_{Ac}$ for each class label and numerical attribute.

$\varepsilon' \leftarrow \frac{\varepsilon}{1+\#\{\text{categorical } A\}+2\#\{\text{numerical } A\}}$
for each class label c **do**
  $\tilde{n}_c \leftarrow L^+(n_c, \varepsilon')$
  $p(c) \leftarrow \frac{\tilde{n}_c}{n}$
  for each categorical A, each value $v \in A$ **do**
    $\tilde{n}_{Avc} \leftarrow L^+(n_{Avc}, \varepsilon')$
    $p(x_A = v|c) = \frac{\tilde{n}_{Avc}}{\tilde{n}_c}$
  for each numerical A **do**
    $\tilde{\mu}_{Ac} \leftarrow L(\mu_{Ac}, \varepsilon')$
    $\tilde{\sigma}_{Ac} \leftarrow L^+(\sigma_{Ac}, \varepsilon')$

of privacy. The anonymized table of data is then shared to the data analyzer to further construct a classifier on this dataset.

5.2.2 Classifiers Over Differentially Private Data

ε-**DP Naïve Bayes classifier**: We use the implementation of ε-DP Naïve Bayes classifier presented in [36]. Algorithm 1 details the whole process and it satisfies ε-DP as proved in [36]. Naïve Bayes relies on the dataset via the queries n_{Avc}, n_c, μ_{Ac} and σ_{Ac} for all $A_c \in \mathcal{A}$. For inserting DP, we use the noisy versions of these values instead as below

$$L^+(n_{Avc}, \varepsilon'), L^+(n_c, \varepsilon'), L^+(\sigma_{Ac}, \varepsilon'), L(\mu_{Ac}, \varepsilon'), \tag{17}$$

where ε' is chosen such that the collection of whole noisy answers satisfies ε-DP. More concretely,

$$\varepsilon' = \frac{\varepsilon}{1 + \#\{\text{categorical } A\} + 2\#\{\text{numerical } A\}}. \tag{18}$$

Note that in (17) we use L^+ for the counts and the standard deviation because they are assumed to be positive, and L for the mean because it has no such restriction. To calculate the expressions in Eq. (17), we need to know their sensitivities. The sensitivities of the counts n_{Avc} and n_c satisfy $\bar{s} = 1$. If for each numerical attribute A, a lower bound l_A and upper bound u_A are public knowledge, the sensitivity of μ_{Ac} and σ_{Ac}, respectively, are given as follows

$$s_{\mu_{Ac}} = \frac{u_A - l_A}{n_c} \quad , \quad s_{\sigma_{Ac}} = \frac{u_A - l_A}{\sqrt{n_c}}. \tag{19}$$

Algorithm 2 Construction of ε-DP SVM classifier

Require: Privacy budget ε; Huber parameter h; overfitting parameter Λ.
Ensure: Separating hyperplane $\bar{w}_{\text{priv}}$.
  $\varepsilon' \leftarrow \varepsilon - \log\left(1 + \frac{1}{nh\Lambda} + \frac{1}{4n^2h^2\Lambda^2}\right)$
  if $\varepsilon' > 0$ **then**
    $\varepsilon'' \leftarrow \varepsilon'$
    $\Lambda' \leftarrow \Lambda$
  else
    $\varepsilon'' \leftarrow \frac{\varepsilon}{2}$
    $\Lambda' \leftarrow \Lambda$
  draw $\bar{b}$ according to $p(\bar{b} = \bar{z}) \propto \mathrm{e}^{-\frac{\varepsilon''}{2}||\bar{z}||}$
  $\bar{w}_{\text{priv}} \leftarrow \arg\max_{\bar{w}} \frac{1}{n}\sum_{i=1}^{n} l_{\text{Huber}}(\bar{w} \cdot \bar{x}^i, x_c^i) + \frac{\Lambda'}{2}||\bar{w}||^2 + \frac{1}{n}\bar{b} \cdot \bar{w}$

ε-DP SVM classifier: We adopt the ε-DP implementation of SVM introduced in [11, 36], which is detailed in Algorithm 2. In SVM, the resulting hyperplane $\bar{w}$ can leak information about D, since it minimizes an objective function J depending on D. To avoid this, the objective function is perturbed so that it does not rely on any row in D significantly. More precisely, instead of the objective function J from Eq. 12, we use:

$$J_{\text{priv}}(\bar{w}, D) = \frac{1}{n}\sum_{i=1}^{n} l_h(\bar{w} \cdot \bar{x}_i, x_i^c) + \frac{\Lambda'}{2}||\bar{w}||^2 + \frac{1}{n}\bar{b} \cdot \bar{w} \tag{20}$$

where $\bar{b}$ is a random vector, whose probability distribution is defined below, and Λ' depends on the choice of Λ and the privacy budget ε. More concretely, given ε, Λ and the Huber parameter h, we define

$$\varepsilon' = \varepsilon - \log\left(1 + \frac{1}{nh\Lambda} + \frac{1}{4n^2h^2\Lambda^2}\right), \tag{21}$$

$$\varepsilon'' = \begin{cases} \varepsilon', & \text{if } \varepsilon' > 0 \\ \frac{\varepsilon}{2}, & \text{otherwise,} \end{cases} \tag{22}$$

$$\Lambda' = \begin{cases} \Lambda, & \text{if } \varepsilon' > 0 \\ \frac{1}{2nh(\mathrm{e}^{\varepsilon/4}-1)}, & \text{otherwise,} \end{cases} \tag{23}$$

and $\bar{b}$ is drawn according to $\mathbb{P}(\bar{b} = \bar{z}) \propto \mathrm{e}^{-\frac{\varepsilon''}{2}||\bar{z}||}$. The algorithm then outputs the hyperplane as below:

$$\bar{w}_{\text{priv}} = \arg\min_{\bar{w}} J_{\text{priv}}(\bar{w}, D). \tag{24}$$

5.2.3 Classifiers Over Encrypted Data

Naïve Bayes classifier over encrypted data: To securely construct a Naïve Bayes classifier over secret shares of data, these probabilities should be computed for all possible attribute-values and class labels [53]. This requires the secure computation of the following counts on the whole distributed data: n (the total number of records); n_{Avc} (the total number of records labeled class C in with the value v of attribute A); n_c (the total number of records with class c).

To compute these values securely, first the data owner P_s $(1 \leq s \leq N)$ locally computes: n^s_{Avc} (the number of records in D_s labeled with class c with the value v of attribute A); n^s_c (the number of records in D_s with class c); n^s (the number of records in D_s). Then, P_s randomly shares one part of n^s_{Avc}, n^s_c, and n^s with server, i.e., $\langle n^s_{Avc}\rangle_1$, $\langle n^s_c\rangle_1$, $\langle n^s\rangle_1$, and the other part with client, i.e., $\langle n^s_{Avc}\rangle_2$, $\langle n^s_c\rangle_2$, $\langle n^s\rangle_2$, such that:

$$\langle n^s_{Avc}\rangle_1 + \langle n^s_{Avc}\rangle_2 \equiv n^s_{Avc} \mod 2^\varsigma \tag{25}$$

$$\langle n^s_c\rangle_1 + \langle n^s_c\rangle_2 \equiv n^s_c \mod 2^\varsigma \tag{26}$$

$$\langle n^s\rangle_1 + \langle n^s\rangle_2 \equiv n^s \mod 2^\varsigma \tag{27}$$

Server and client shuffle their data and then employ addition gates to obtain (on the base of $\mod 2^\varsigma$):

$$\langle n_{Avc}\rangle = \langle n^1_{Avc}\rangle_1 + \cdots + \langle n^N_{Avc}\rangle_1 + \langle n^1_{Avc}\rangle_2 + \cdots + \langle n^N_{Avc}\rangle_2 \tag{28}$$

$$\langle n_c\rangle = \langle n^1_c\rangle_1 + \cdots + \langle n^N_c\rangle_1 + \langle n^1_c\rangle_2 + \cdots + \langle n^N_c\rangle_2 \tag{29}$$

$$\langle n\rangle = \langle n^1\rangle_1 + \cdots + \langle n^N\rangle_1 + \langle n^1\rangle_2 + \cdots + \langle n^N\rangle_2 \tag{30}$$

From these values, server and client can compute the building blocks of the Naïve Bayes classifier as $p(c) = \frac{n_c}{n}$ and $p(x_A = v|c) = \frac{n_{Avc}}{n_c}$ for all v, c, A. These values are then shared with the data owners. Algorithm 3 summarizes the whole process.

SVM over encrypted data: To construct an SVM classifier over secret shares of data, it requires to compute $x^c_i \gamma_i k(\mathbf{x_i}, \mathbf{x_j}) x^c_j \gamma_j$ for all i, j, where $\mathbf{x_i}$, $\mathbf{x_j}$, x^c_i, and x^c_j are given in protected format (secret shares), and γ_i and γ_j should optimally be computed solving the optimization problem presented in Eq. 15 [53]. Also, we assume that the kernel function is linear, i.e., $k(\mathbf{x_i}, \mathbf{x_j}) = \mathbf{x_i} \cdot \mathbf{x_j}$. To this end, for every record $\mathbf{x^s} = (x^s_1, \ldots, x^s_m, x^{cs})$ in D_s, the data owner P_s randomly generates the shares:

$$\langle \mathbf{x^s}\rangle_1 = (\langle x^s_1\rangle_1, \ldots, \langle x^s_m\rangle_1, \langle x^{cs}\rangle_1) \tag{31}$$

$$\langle \mathbf{x^s}\rangle_2 = (\langle x^s_1\rangle_2, \ldots, \langle x^s_m\rangle_2, \langle x^{cs}\rangle_2) \tag{32}$$

such that $\langle x^s_i\rangle_1 + \langle x^s_i\rangle_2 \equiv x^s_i \mod 2^\varsigma$ and $\langle x^{cs}\rangle_1 + \langle x^{cs}\rangle_2 \equiv x^{cs} \mod 2^\varsigma$. Then, P_s sends $\langle \mathbf{x^s}\rangle_1$ and $\langle \mathbf{x^s}\rangle_2$ to server and client, respectively. After collecting the shares of all records, server and client use addition and multiplication gates to compute:

Algorithm 3 Construction of encryption-based Naïve Bayes Classifier

Require: Datasets $D_1, \ldots, D_N$ (horizontally) owned by $P_1, \ldots, P_N$, respectively.
Ensure: Naïve Bayes classifier constructed over $D_1 \cup \ldots \cup D_N$.
for $1 \leq s \leq N$ **do**
  P_s locally computes:
    n^s_{Avc}: the number of records in D_s with class c with value v of attribute A.
    n^s_c: the number of records in D_s labeled c.
    n^s: the number of records in D_s.
  P_s shares:
    $\langle n^s_{Avc}\rangle_1, \langle n^s_c\rangle_1, \langle n^s\rangle_1$ with the server and $\langle n^s_{Avc}\rangle_2, \langle n^s_c\rangle_2, \langle n^s\rangle_2$ with the client.
for $P_s,\ A \in \mathcal{A},\ v \in A,\ c \in C$ **do**
  server and client using addition gate compute:
    $\langle n_{Avc}\rangle = \langle n^1_{Avc}\rangle_1 + \ldots + \langle n^N_{Avc}\rangle_1 + \langle n^1_{Avc}\rangle_2 + \ldots + \langle n^N_{Avc}\rangle_2 \mod 2^\varsigma$
    $\langle n_c\rangle = \langle n^1_c\rangle_1 + \ldots + \langle n^N_c\rangle_1 + \langle n^1_c\rangle_2 + \ldots + \langle n^N_c\rangle_2 \mod 2^\varsigma$
    $\langle n\rangle = \langle n^1\rangle_1 + \ldots + \langle n^N\rangle_1 + \langle n^1\rangle_2 + \ldots + \langle n^N\rangle_2 \mod 2^\varsigma$
    return $p(c) = \frac{n_c}{n}$ and $p(x_A = v|c) = \frac{n_{Avc}}{n_c}$

Algorithm 4 Construction of encryption-based SVM Classifier

Require: Datasets $D_1, \ldots, D_N$ (horizontally) owned by $P_1, \ldots, P_N$, respectively.
Ensure: SVM classifier constructed over $D_1 \cup \ldots \cup D_N$.
for $\mathbf{x^s} = (x^s_1, \ldots, x^s_m, x^{cs}) \in D_s$ **do**
  P_s shares:
    $\langle \mathbf{x^s}\rangle_1 = (\langle x^s_1\rangle_1, \ldots, \langle x^s_m\rangle_1, \langle x^{cs}\rangle_1)$ with server
    $\langle \mathbf{x^s}\rangle_2 = (\langle x^s_1\rangle_2, \ldots, \langle x^s_m\rangle_2, \langle x^{cs}\rangle_2)$ with client
for $\mathbf{x^s_i}, \mathbf{x^{s'}_j} \in \mathcal{D}\ \ \forall 1 \leq s, s' \leq N$ **do**
  server and client using addition and multiplication gates compute:
    $\langle k(\mathbf{x^s_i}, \mathbf{x^{s'}_j})\rangle \ = \langle x^s_{i1}\rangle \times \langle x^{s'}_{j1}\rangle + \ldots + \langle x^s_{im}\rangle \times \langle x^{s'}_{jm}\rangle$
    $\langle x^c_i \times x^c_j\rangle = \langle x^c_i\rangle \times \langle x^c_j\rangle$
    return $k(\mathbf{x^s_i}, \mathbf{x^{s'}_j}) \cdot x^{cs}_i \cdot x^{cs}_j$

$$\langle k(\mathbf{x}_i, \mathbf{x}_j)\rangle = \langle x_{i1}\rangle \times \langle x_{j1}\rangle + \cdots + \langle x_{im}\rangle \times \langle x_{jm}\rangle \tag{33}$$

$$\langle x^c_i \times x^c_j\rangle = \langle x^c_i\rangle \times \langle x^c_j\rangle \tag{34}$$

where $\times$ refers to the multiplication protocol defined over secret shares. The values $\langle k(\mathbf{x}_i, \mathbf{x}_j)\rangle$ and $\langle x^c_i \times x^c_j\rangle$ are then used to solve the optimization problem (Eq. 15) to build the SVM classifier. The final output is shared with all data owners. Algorithm 4 summarizes this process.

5.3 Experimental Analysis

This section compares the implementation of Naïve Bayes and SVM classifiers in anonymized, differentially private, and encrypted setting on Adult dataset. Adult dataset contains 14 attributes (both numerical and categorical) such as *age*, *occupa-*

Table 2 The accuracy of Naïve Bayes and SVM classifiers on anonymized data

		3-anonymity		2-diversity		0.2-closeness	
Classifier	Original	Accuracy	Ratio	Accuracy	Ratio	Accuracy	Ratio
Naïve Bayes	71.87	59.23	82.33	56.71	78.90	37.90	52.73
SVM	82.93	73.30	88.38	66.04	79.63	51.44	62.02

tion, *education*, *working class*, etc. These attributes are used to predict the income of an individual, which has two possible values '$>50K$' and '$<50K$'. This dataset contains 48842 instances.[5]

5.3.1 Anonymized Classifier Results

The accuracy of Naïve Bayes and SVM classifiers trained on original Adult dataset and its anonymized versions, i.e., 3-anonymous, 2-diverse, and 0.2-closeness dataset, are presented in Table 2. The results indicate that, in general, SVM classifier produces more accurate classification result compared to Naïve Bayes classifier on this dataset.

To gain better insight on the impact of anonymization techniques on these classifiers, we compare the ratio values, which measures the accuracy of the classifiers trained on original data divided to the accuracy of the classifier trained on anonymized data. The higher ratio values for SVM classifier demonstrate that this classifier slightly outperforms the Naïve Bayes classifier in pertaining the accuracy close to the accuracy of original data. This can be resulted from the fact that the anonymization process makes the datasets linearly separable resulting in (generally) higher performance for the SVM classifier compared to the Naïve Bayes classifier.

5.3.2 Differentially Private Classifier Results

The performance evaluation of the Naïve Bayes and SVM classifiers in non-private setting and ε-DP setting (Algorithms 1 and 2) has been implemented in Python [36]. The privacy budgets ε used to train the classifiers in the ε-DP setting are taken from the set $\mathcal{E} = \{10^{-11}, 0.001, 0.005, 0.01, 0.05, 0.1, 0.25, 0.5, 0.75, 1\}$. The results on accuracy are shown in Fig. 6. It can be observed that the SVM classifier is slightly more accurate than Naïve Bayes when trained in ε-DP setting on Adult dataset. This is mainly resulted from the high number of records of this dataset, as the SVM structure is shaped based on a hyperplane determining by the support vectors' distances. When the dataset comprises a large number of records, the noises added through DP negligibly affect the hyperplane location [36].

[5] https://archive.ics.uci.edu/ml/datasets/adult.

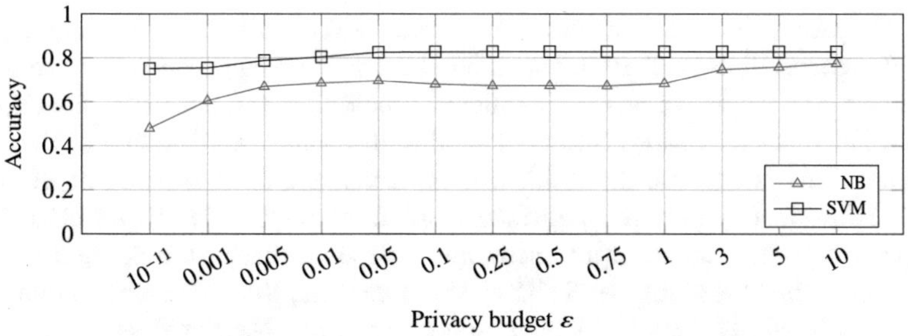

Fig. 6 DP-based classifiers accuracy

Table 3 Naïve Bayes and SVM classifiers accuracy in ε-DP setting

		$\varepsilon = 0.1$		$\varepsilon = 1$		$\varepsilon = 5$	
Classifier	Original	Accuracy	Ratio	Accuracy	Ratio	Accuracy	Ratio
Naïve Bayes	71.87	68.48	95.28	68.58	95.42	70.45	98.02
SVM	82.93	82.78	99.81	82.88	99.93	82.88	99.93

Table 4 Computation and communication costs of Naïve Bayes and SVM classifiers trained on encrypted data

Classifier	Accuracy	Execution time (ms)	Bandwidth usage (bytes)
Naïve Bayes	71.87	21.90	979,764
SVM	82.93	265,917.52	174,098,320

Table 3 reports the ratio and exact accuracy of ε-DP classifiers for some selected privacy budgets. As can be observed it confirms that the SVM classifier slightly outperforms the Naïve Bayes classifier for Adult dataset both in terms of accuracy and ratio values.

5.3.3 Encrypted Classifier Results

The Naïve Bayes and SVM classifiers have been trained in encrypted-based setting using 32-bit Arithmetic sharing in ABY framework (i.e., $\zeta = 32$). The experiments are performed on a single machine running Ubuntu 18.04 LTS with a 64-bit microprocessor and 16 GB of RAM, with Intel Core i7-4770, 3.40 GHz x 8 [53]. The total computation and communication costs of training Naïve Bayes and SVM classifiers on Adult dataset are shown in Table 4 (for four data providers). The bandwidth usage is the summation of sent and received bytes. The total execution time also includes the time of setup and online phases [13].

A comparison between the Naïve Bayes and SVM classifiers shows that the accuracy of two classifiers is almost comparable. However, we can observe that the secure construction of the Naïve Bayes classifier is almost 12,000 times faster and requires almost 177 times less bandwidth compared to the secure construction of the SVM classifier. This is because the *Adult* dataset contains a large number of instances ($\sim$50K) described with only 14 attributes and training the Naïve Bayes classifier is irrespective of the number of instances, and it depends on the number of attributes. On the other hand, training an SVM classifier exhibits the high computation costs for the *Adult* dataset due to its large number of records. Nevertheless, the accuracy of the SVM classifier is slightly better than the accuracy of the Naïve Bayes classifier for this dataset. This higher accuracy might (mis)lead the data owners into selecting SVM rather than Naïve Bayes, while SVM might not be suitable due to application-domain bandwidth limitations and computation constraints.

Our experimental results for these two classifiers confirms that a classifier might outperform the other one under specific conditions. In the following section, we present the limitations and benefits of privacy models in private classifiers construction.

6 Discussion

In this section, the significant findings on the usage of different privacy models in constructing the private classifiers is discussed. Furthermore, the main attack models against the training dataset and against the private classifier models are explained.

6.1 Findings

Here we summarize all findings (from our experiments and other studies) on employing anonymization, DP, and encryption techniques in constructing private classifiers.

6.1.1 Anonymization for Classifiers Construction

- The anonymization approaches remove information from dataset by generalization or suppression, which results in degrading the classification performance. Different classifiers show various sensitivity to such information loss and there is not a single classifier that outperforms other classifiers for all performance metrics [55]. Nonetheless, it has been observed that the classifiers with tendency to separate data linearly (e.g., Linear Regression and SVM) demonstrate better performance in classifying anonymized data [5].

- The selection of features for generalization plays a key role in classifiers performance over anonymized data. The generalization of features which are strongly correlated to class labels have considerably negative impact on the classifiers performance [55]. Accordingly, a feature ranking technique can guide the data provider to select which features are prioritized to be generalized.
- It has been found that among dataset properties, the number of class labels considerably affects the classifiers' performance based on the anonymized dataset, i.e., the datasets with only two class labels are a better source for training the classifiers over the anonymized data [5]. This suggests the division of a dataset with more than two class labels into several datasets with only two class labels. The dataset size and the number of attributes have negligible (or no) impact on classifiers performance.
- The variation of anonymization parameters, i.e., the values of k, ℓ, and t, does not affect the *trend* of classifiers' performance over anonymized datasets- apart from some exceptions with a negligible difference [5].

6.1.2 Differential Privacy for Classifiers Construction

- DP-based approaches are defined based on data perturbation such that an adversary with the background knowledge about data is still incapable of linking an individual record to a dataset. One of the big challenges in the application of DP for classification is finding the best trade-off between privacy and model accuracy. In DP-based classifiers, not only the selection of privacy budget affects this trade-off score, but also the classifiers' hyper parameters have direct impact on this outcome. This is mainly resulted from the inherent property of DP in which noise should be inserted in each query for building the model. For instance, for constructing a Random Forest in DP setting, the privacy budget, the maximum tree depth, and the number of trees in Random Forest affect the classifiers' privacy and accuracy [23]. This means that the classifiers with the lower number of hyper parameters can better serve in finding the best trade-off score.
- It has been shown that none of the well-known classification algorithms is a *one-size-fits-all* solution for all datasets and privacy budgets [36]. Still, some classifiers might outperform the other ones under some specific conditions. For instance, the SVM classifier outperforms the Naïve Bayes classifier for a large dataset.
- It has been observed that the randomness property of DP leads to a non-smooth behaviour of private classifiers in terms of accuracy for different ε values [36]. This results in observing that the accuracy of a private classifier remains unchanged when the privacy budget ε varies in a specific range of values, or even it becomes better for more restricted privacy constrain.
- While the computation overhead is generally considered as an obstacle in the application of encryption techniques for building private classifiers, it has been found that the construction of DP-based classifiers can also be quite heavy. The reason for this is that, differently from anonymization, some (it could be many)

communications between data provider and data analyzer should be executed, while in each round of communication appropriate noise should be added to the query [67].
- The classifiers trained under DP are more resistant against most attacks as the DP blurs the information to reduce the privacy risk (even against an attacker with prior knowledge) at the cost of utility loss [34].

6.1.3 Encryption for Classifiers Construction

- The application of encryption techniques in building classifiers imposes computation (runtime) and communication (bandwidth usage) overheads.
- The aforementioned costs are specifically noticeable when the classifiers structure requires to be learnt pointwise (e.g., SVM and kNN [68]) on a large dataset, or requires to be learnt through an iterative process (e.g., in deep learning [54]). This outcome suggests the implementation of probability-based classifiers, e.g., Naïve Bayes classifier, where the classifier has to be trained over constrained resources, even if their accuracy might be (slightly) lower than the accuracy of other classifiers for one dataset [53].
- Encryption-based techniques can protect the intermediate steps of computation against attacks, when the collaborative classifier is trained on the distributed data [54].

6.2 Security Risks

To provide insight on the security guaranteed by different privacy models in constructing private classifiers, we define the treat/attack models and then compare the resistance of privacy models against these attacks. These attack models are presented in two main categories: (i) privacy of the published dataset and (ii) privacy of the classifier model.

6.2.1 Privacy of Published Dataset

In some scenarios, the data provider perturbs a dataset and then shares it with data analyzer. This is generally a valid operation in employing anonymization and differential privacy (in releasing a table of data). The attacks threatening this published dataset are the followings.

Record/Attribute/Table linkage attacks: A linkage attack includes *record linkage*, *attribute linkage*, and *table linkage* attacks, which respectively happen when an adversary is able to link a data owner to a record in a published dataset, link a

Table 5 Comparison of attack models on the published dataset

Privacy model	Record linkage	Attribute linkage	Table linkage	Probabilistic
k-anonymity	✓	x	x	x
ℓ-diversity	✓	✓	x	x
t-closeness	x	✓	x	✓
Differential privacy	x	x	✓	✓

data owner to a SA in a published dataset, or link a data owner to the published table of data by itself [61]. Linkage attacks are distinguished by the adversary's prior knowledge about QIs. In linkage and attribute linkage, the adversary knows about the presence of a specific individual's data in a dataset and wants to learn about the sensitive information of that particular data owner. However, Table linkage attack tries to understand whether a known individual's information is available in the published dataset. Moreover, in an attribute linkage attack, the adversary could gather sensitive information from the published dataset based on the distribution of sensitive values in the group to which the individual belongs.

Probabilistic attack: Probabilistic attack happens when an attacker changes his probabilistic belief about the sensitive information of a data owner after accessing the published data. In general, this subset of privacy models seeks to achieve the uninformative principle, which aims to keep the gap between pre- and posterior beliefs as minimal as possible.

Table 5 summarizes the privacy models applicable on published dataset and the attacks to which they are resistant against.

6.2.2 Privacy of Classifiers' Model

In this section, we introduce the main attack models against the private classifier model and compare the capability of privacy preserving techniques in defending against these attacks [34].

Model extraction attack: The aim of the model extraction attack is the duplication (i.e., stealing) a model of the classifier model (AI model in general).

Feature estimation attack: A feature estimation attack aims to estimate the presence of certain features in the training dataset.

Membership inference attack: This attack aims to acquire knowledge about whether a certain data record has been presented in model's training dataset or not.

Model memorization attack: The model memorization attack targets recovering the exact feature values on an individual's samples.

In general, DP can work effectively against membership inference attack by definition, as DP is designed to make the data owners' records indistinguishable. It is also

Table 6 Comparison of attack models on the classifier's model

Privacy model	Model extraction	Feature estimation	Membership inference	Model memorization
Differential privacy	✓	x	✓	x
Encryption	x	✓	x	✓

resistant against model extraction attacks when noise is added to the model learning parameters. Encryption, on the other hand, can maintain the adversary's knowledge to a black-box case, thus it is effective to white-box attacks like model memorization and feature estimation attacks. Table 6 summarizes the privacy models and the attacks that each one is resistant against [34].

7 Conclusion

Machine learning models require a lot of datasets and details, some of which may contain sensitive or personal information, to achieve reliable levels of predictive performance. A privacy preserving technique should be then incorporated into classifiers to ensure data is adequately protected in accordance with current data protection regulations, and thereby reduce the risk of unauthorized data access. In this chapter, we investigated three well-known privacy preserving models to privatize data that is used in the classifier construction. The comparison of private Naïve Bayes and SVM classifiers building over the private inputs is analyzed in terms of accuracy and training construction costs. Detailed results are provided indicating how privacy models can be used to build classifiers in private setting. Finally, attacks focused on the training data and private classifier models are discussed. The result of this work provides an insight on the impact of different criteria on the performance of private classifiers' performance.

References

1. Aggarwal CC (2014) Data classification: algorithms and applications. Chapman and Hall CRC
2. Agrawal D, Aggarwal CC (2001) On the design and quantification of privacy preserving data mining algorithms. In: Symposium on principles of database systems, pp 247–255
3. Ahmed A, Krishnan VVG, Foroutan SA, Touhiduzzaman M, Rublein C, Srivastava A, Wu Y, Hahn A, Suresh S (2019) Cyber physical security analytics for anomalies in transmission protection systems. IEEE Trans Ind Appl 55(6):6313–6323
4. AlHussaeni K, Fung BCM, Iqbal F, Liu J, Hung PCK (2018) Differentially private multidimensional data publishing, pp 717–752

5. Alishahi M, Zannone N (2021) Not a free lunch, but a cheap one: on classifiers performance on anonymized datasets. In: Data and applications security and privacy conference (DBSec). Lecture notes in computer science, vol 12840. Springer, pp 237–258
6. Aono Y, Hayashi T, Phong LT, Wang L (2016) Scalable and secure logistic regression via homomorphic encryption. In: Conference on data and application security and privacy, pp 142–144 (2016)
7. Blum A, Dwork C, McSherry F, Nissim K (2005) Practical privacy: the SuLQ framework. In: International conference on principles of database systems. ACM, pp 128–138
8. Bost R, Popa R, Tu S, Goldwasser S (2014) Machine learning classification over encrypted data. IACR Cryptol. ePrint Arch 2014:331
9. Brickell J, Shmatikov V (2008) The cost of privacy: destruction of data-mining utility in anonymized data publishing. In: International conference on knowledge discovery and data mining. ACM, pp 70–78
10. Bünz B, Agrawal S, Zamani M, Boneh D (2020) Zether: towards privacy in a smart contract world. In: Financial cryptography and data security, pp 423–443
11. Chaudhuri K, Monteleoni C, Sarwate AD (2011) Differentially private empirical risk minimization. J Mach Learn Res 12(29):1069–1109
12. Ciriani V, di Vimercati SDC, Foresti S, Samarati P (2008) k-anonymous data mining: a survey. In: Privacy-preserving data mining: models and algorithms, pp 105–136
13. Demmler D, Schneider T, Zohner M (2015) ABY—a framework for efficient mixed-protocol secure two-party computation. In: Annual network and distributed system security symposium. Internet Society
14. Dwork C, McSherry F, Nissim K, Smith A (2006) Calibrating noise to sensitivity in private data analysis. In: Theory of cryptography. Springer, pp 265–284
15. Fan W, He J, Guo M, Li P, Han Z, Wang R (2020) Privacy preserving classification on local differential privacy in data centers. J Parallel Distrib Comput 135:70–82
16. Fathalizadeh A, Moghtadaiee V, Alishahi M (2022) On the privacy protection of indoor location dataset using anonymization. Comput Secur
17. Fletcher S, Islam MZ (2019) Decision tree classification with differential privacy: a survey. ACM Comput Surv 52(4):1–33
18. Friedman A, Schuster A, Wolff R (2006) k-anonymous decision tree induction. In: Knowledge discovery in databases, pp 151–162
19. Gao C, Li J, Xia S, Choo KR, Lou W, Dong C (2020) Mas-encryption and its applications in privacy-preserving classifiers. IEEE Trans Knowl Data Eng 1–17
20. Gati NJ, Yang LT, Feng J, Nie X, Ren Z, Tarus SK (2021) Differentially private data fusion and deep learning framework for cyber-physical-social systems: state-of-the-art and perspectives. Inf Fusion 76:298–314
21. Gong M, Xie Y, Pan K, Feng K, Qin A (2020) A survey on differentially private machine learning. IEEE Comp Intell Mag 15(2):49–64
22. Gursoy ME, Inan A, Nergiz ME, Saygin Y (2017) Differentially private nearest neighbor classification. Data Min Knowl Discov 31(5):1544–1575
23. Hou J, Li Q, Meng S, Ni Z, Chen Y, Liu Y (2019) Dprf: a differential privacy protection random forest. IEEE Access 7:130707–130720. https://doi.org/10.1109/ACCESS.2019.2939891
24. Inan A, Kantarcioglu M, Bertino E (2009) Using anonymized data for classification. In: International conference on data engineering, pp 429–440
25. Jagannathan G, Pillaipakkamnatt K, Wright RN (2009) A practical differentially private random decision tree classifier. In: International conference on data mining. IEEE, pp 114–121
26. Jain P, Thakurta A (2013) Differentially private learning with kernels. In: International conference on machine learning, pp 118–126
27. Jayaraman B, Evans D (2019) Evaluating differentially private machine learning in practice. In: USENIX conference on security symposium, SEC'19, pp 1895–1912
28. Khodaparast F, Sheikhalishahi M, Haghighi H, Martinelli F (2018) Privacy preserving random decision tree classification over horizontally and vertically partitioned data. In: International conference on dependable, autonomic and secure computing. IEEE, pp 600–607

29. Khodaparast F, Sheikhalishahi M, Haghighi H, Martinelli F (2019) Privacy-preserving LDA classification over horizontally distributed data. In: International symposium on intelligent distributed computing, pp 65–74
30. Li N, Li T, Venkatasubramanian S (2007) t-closeness: privacy beyond k-anonymity and l-diversity. In: 23rd international conference on data engineering. IEEE, pp 106–115
31. Li T, Li J, Liu Z, Li P, Jia C (2018) Differentially private naive bayes learning over multiple data sources. Inf Sci 89–104
32. Li T, Li N, Zhang J, Molloy I (2012) Slicing: a new approach for privacy preserving data publishing. IEEE Trans Knowl Data Eng 24(3):561–574. https://doi.org/10.1109/TKDE.2010.236
33. Lin KP, Chen MS (2011) On the design and analysis of the privacy-preserving SVM classifier. IEEE Trans Knowl Data Eng 1704–1717
34. Liu B, Ding M, Shaham S, Rahayu W, Farokhi F, Lin Z (2020) When machine learning meets privacy: a survey and outlook. arXiv:2011.11819
35. Liu X, Lu R, Ma J, Chen L, Qin B (2016) Privacy-preserving patient-centric clinical decision support system on naïve bayesian classification. IEEE J Biomed Health Inf 20(2):655–668
36. Lopuhaä-Zwakenberg M, Alishahi M, Kivits J, Klarenbeek J, van der Velde GJ, Zannone N (2021) Comparing classifiers' performance under differential privacy. In: International conference on security and cryptography (SECRYPT)
37. Machanavajjhala A, Kifer D, Gehrke J, Venkitasubramaniam M (2007) l-diversity: privacy beyond k-anonymity. ACM Trans Knowl Discov Data 1(1):3–es
38. Malle B, Kieseberg P, Holzinger A (2017) Do not disturb? Classifier behavior on perturbed datasets. In: Machine learning and knowledge extraction, pp 155–173
39. Malle B, Kieseberg P, Weippl E, Holzinger A (2016) The right to be forgotten: towards machine learning on perturbed knowledge bases. In: Availability, reliability, and security in information systems, pp 251–266
40. Martinelli F, Alishahi MS (2019) Distributed data anonymization. In: Conference on dependable, autonomic and secure computing (DASC), pp 580–586
41. McDonald AWE, Afroz S, Caliskan A, Stolerman A, Greenstadt R (2012) Use fewer instances of the letter "i": toward writing style anonymization. In: Privacy enhancing technologies, pp 299–318
42. Mishra P, Lehmkuhl R, Srinivasan A, Zheng W, Popa R (2020) Delphi a cryptographic inference service for neural networks. In: USENIX security symposium. USENIX Association, pp 2505–2522
43. Naehrig M, Lauter K, Vaikuntanathan V (2011) Can homomorphic encryption be practical? In: Cloud computing security workshop. ACM, pp 113–124
44. Nergiz ME, Gök MZ (2014) Hybrid k-anonymity. Comput Secur 44:51–63
45. Prince P, Lovesum S (2020) Privacy enforced access control model for secured data handling in cloud-based pervasive health care system. SN Comput Sci 1(239)
46. Rubinstein BIP, Bartlett PL, Huang L, Taft N (2009) Learning in a large function space: privacy-preserving mechanisms for SVM learning. CoRR, abs/0911.5708
47. Samarati P (2001) Protecting respondents' identities in microdata release. IEEE Trans Knowl Data Eng 13(6):1010–1027
48. Sazdar AM, Ghorashi SA, Moghtadaiee V, Khonsari A, Windridge D (2020) A low-complexity trajectory privacy preservation approach for indoor fingerprinting positioning systems. J Inf Secur Appl 53:1–9
49. Semwal P, Handa A (2022) Cyber-attack detection in cyber-physical systems using supervised machine learning. Springer International Publishing, Cham, pp 131–140. https://doi.org/10.1007/978-3-030-74753-4_9
50. Shaukat K, Luo S, Varadharajan V, Hameed IA, Xu M (2020) A survey on machine learning techniques for cyber security in the last decade. IEEE Access 8:222310–222354
51. Sheikhalishahi M, Martinelli F (2017) Privacy-utility feature selection as a privacy mechanism in collaborative data classification. In: Enabling technologies: infrastructure for collaborative enterprises, pp 244–249

52. Sheikhalishahi M, Saracino A, Martinelli F, Marra AL (2021) Privacy preserving data sharing and analysis for edge-based architectures. Int J Inf Secur 1(2):1–23
53. Sheikhalishahi M, Zannone N (2020) On the comparison of classifiers' construction over private inputs. In: International conference on trust, security and privacy in computing and communications. IEEE, pp 691–698
54. Shokri R, Shmatikov V (2015) Privacy-preserving deep learning. In: ACM SIGSAC conference on computer and communications security, CCS '15. Association for Computing Machinery, pp 1310–1321
55. Slijepcevic D, Henzl M, Klausner LD, Dam T, Kieseberg P, Zeppelzauer M (2021) k-anonymity in practice: how generalisation and suppression affect machine learning classifiers. Comput Secur 111:102488
56. Sweeney L (2002) k-anonymity: a model for protecting privacy. Int J Uncertainty Fuzziness Knowl Based Syst 10(05):557–570
57. Vaidya J, Clifton C, Kantarcioglu M, Patterson AS (2008) Privacy-preserving decision trees over vertically partitioned data. ACM Trans Knowl Discov Data 2(3)
58. Vaidya J, Kantarcıoğlu M, Clifton C (2008) Privacy-preserving naïve bayes classification. VLDB J 17(4)
59. Vaidya J, Shafiq B, Basu A, Hong Y (2013) Differentially private naive bayes classification. International joint conferences on web intelligence and intelligent agent technologies 1:571–576
60. Xiao M, Han K, Huang L, Li J, Privacy preserving C4.5 algorithm over horizontally partitioned data. In: International conference on grid and cooperative computing, pp 78–85 (2006)
61. Xu Y, MaMeili T, Tian T (2014) A survey of privacy preserving data publishing using generalization and suppression. Appl Math Inf Sci 8(3):1103–1116
62. Yang Q, Liu Y, Chen T, Tong Y (2019) Federated machine learning: concept and applications. ACM Trans Intell Syst Technol 10(2)
63. Ye M, Wu X, Hu X, Hu D (2013) Anonymizing classification data using rough set theory. Knowl Based Syst 43
64. Yilmaz E, Al-Rubaie M, Chang JM (2019) Locally differentially private naive bayes classification. arXiv:1905.01039
65. Yu H, Vaidya J, Jiang X (2006) Privacy-preserving SVM classification on vertically partitioned data. In: Advances in knowledge discovery and data mining. Springer
66. Zafarani F, Clifton C (2020) Differentially private naïve bayes classifier using smooth sensitivity. arXiv:2003.13955
67. Zhang L, Liu Y, Wang R, Fu X, Lin Q (2017) Efficient privacy-preserving classification construction model with differential privacy technology. J Syst Eng Electr 28(1):170–178. https://doi.org/10.21629/JSEE.2017.01.19
68. Zuber M, Sirdey R (2021) Efficient homomorphic evaluation of k-nn classifiers. Proc Privacy Enhancing Technol 2021:111–129

Usable Identity and Access Management Schemes for Smart Cities

Sandeep Gupta and Bruno Crispo

Abstract Usable Identity and Access Management (IAM) schemes are highly required to control and track users' identity and access privileges for a safe and secure smart city. Any safety or security breach in critical infrastructures, e.g., smart financial solutions, smart transportation, and smart buildings, can disrupt the normal life of its residents. Studies have reported that traditional knowledge- and token-based IAM schemes are unable to fully secure these emerging use cases due to inherent security and usability issues in them. This chapter presents multi-modal biometric-based IAM schemes for smart payment apps, smart transportation, and smart buildings that can partially address the safety and security concerns of residents. We also describe the framework for designing risk-based, implicit, or continuous verification IAM schemes for such use cases.

Keywords Identity and access management · Biometrics · Smart financial solutions · Smart transportation · Smart buildings · Smart city

1 Introduction

The success of a "smart city" extensively depends on smarter and fully secure cyber-physical systems (CPS) for improving people's quality of life. Among various CPS, smart financial solutions, smart transportation, and smart buildings are some of the most important for the sustainability of smart cities. However, any security breach in

S. Gupta (✉) · B. Crispo
Department of Information Engineering & Computer Science (DISI), University of Trento, Trento, Italy
e-mail: sandeep.gupta@ex-staff.unitn.it

B. Crispo
e-mail: bruno.crispo@unitn.it

T. Dimitrakos et al. (eds.), *Collaborative Approaches for Cyber Security in Cyber-Physical Systems*, Advanced Sciences and Technologies for Security Applications,
https://doi.org/10.1007/978-3-031-16088-2_3

these strategic solutions could pose considerable risks to the prospects of smart cities. Verizon's Data Breach Investigations Report (DBIR) [34] analyzed 29,207 real-world security incidents, claiming that 85% of breaches involved a human element, which is a pernicious trend. DBIR further reported that the top three breaches were social engineering (35%), basic web application attacks (24%), and system intrusion (18%).

Clearly, traditional knowledge- and token-based identity and access management schemes are not highly effective to secure emerging CPS [20, 37]. Impersonation-, observation- and brute force-attacks can easily exploit traditional verification mechanisms [13, 23]. Nonetheless, weak passwords remain the major cause of botnet-based attacks like Mirai resulting in Denial of Services (DoS) of CPS [3, 24]. Also, these traditional mechanisms have shown the inability to fulfill the usability requirements of the end-users [6, 33]. Therefore, a thorough investigation for usable identity and access management schemes that can secure smart cities and the underlying smart CPS is inevitable.

This chapter discusses biometric-based IAM schemes [14] for smart payment apps, smart transportation, and smart buildings. In addition to securing critical services, we take this opportunity to reify a safer and more secure smart city. The main contributions of this chapter are as follows.

1. Examples of multi-modal biometric-based IAM schemes for smart payment apps, smart transportation, and smart buildings to address the safety and security concerns of smart city residents.
2. Framework for designing risk-based, implicit, or continuous verification IAM schemes.

The rest of the chapter is organized as follows: Sect. 2 covers the drawbacks in traditional verification schemes and presents the building blocks of a biometric-based identity and access mechanism. Section 4 presents HOLD & TAP user verification scheme for smart payment apps. Section 5 presents DRIVERAUTH risk-based user verification schemes for smart transportation such as on-demand rides strengthening the security and safety of their customers. Section 6 presents STEP & TURN a secure and usable user verification scheme for cyber-physical space to secure access to their authorized users. Finally, Sect. 8 concludes the chapter.

2 Background

The term "smart city" typically connotes an interplay of cyber-physical systems orchestrating smart financial solutions, smart transportation, smart buildings, etc., to deliver the residents a better quality of life. However, considering the substantial increase in security incidents [34], it becomes imperative to redesign IAM schemes that rely on traditional verification mechanisms by introducing biometrics for a secure and safer smart city [30].

2.1 Drawbacks in Traditional Verification Schemes

The biggest drawback of IAM schemes employing traditional verification schemes like Personal Identification Numbers (PINs), passwords, or access cards is that they can give access to anyone knowing the PIN/password or having a genuine access card. Studies have shown that PINs and passwords can be easily guessed, shared, cloned, or stolen [9, 23]. Binbeshr et al. [5] described that PIN-entry methods are highly susceptible to observation attacks. The adversary can observe the login PIN directly or use a recording tool to gain illegitimate access later. Moreover, adversaries can exploit default usernames and passwords of IoT devices for installing botnets or worms to carry out DoS/DDoS attacks on CPS [12, 25].

Traditional verification schemes can also be an easy victim of phishing, impersonation, or insider attacks where the impostor misuses the information of legitimate users to reset the user pin/password or to get a duplicate access card [1, 13, 28]. Nevertheless, with the availability of enormous computation resources, brutal-force attacks can try a large number of users' identity-related information to generate different PINs/passwords, thus, to gain access to CPS [10]. To overcome such shortcomings, the system enforces stringent password policies (e.g., must have uppercase and lowercase letters and special characters), however, that adversely affect the usability of IAM schemes [37].

Service providers adopted two-factor verification, e.g., username/password and One-Time-Passcodes (OTP), to strengthen the security of their online security-sensitive solutions. This further deteriorates the usability because the users are compulsorily required to manage additional hardware to access the services. Moreover, the idea of enhancing security with multi-factor verification perishes too due to side-channel attacks, e.g., MITM (Man-in-the-Middle), and MITPC/Phone (Man-in-the-PC/Phone) [8]. Overall, usability issues such as cognitive load, form-factor, additional hardware, etc., make traditional identity and access systems unsuitable for emerging use-cases.

Zhang et al. [36] presented a study on security and privacy issues that included privacy leakage, secure information processing, and dependability in control. The author emphasizes that any unauthorized access to urban public facilities in a smart city must be prevented. Consequently, IAM schemes require rethinking, with biometrics providing an appropriate alternative to overcoming the drawbacks present in traditional verification schemes.

2.2 Biometric-Based Identity and Access System

Figure 1 shows the basic building blocks of a biometric-based identity and access system that primarily consists of five modules: (1) data acquisition module, (2) data processing module, (3) features processing module, (4) database module, and (5) classification module. The ISO standard:24741 [19] specified the term biometrics

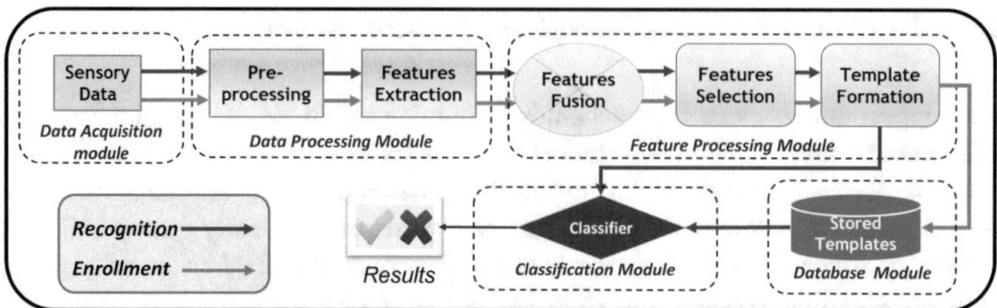

Fig. 1 Biometric-based identity and access system

or biometric recognition as "*the automated recognition of individuals based on their biological and behavioral characteristics*."

1. **Data Acquisition Module**: This module consists of sensors that acquire one or more biometric traits of an individual. It is desired that the measured biometric traits are both distinctive between individuals and repeatable over time for the same individual. This module can also be referred to as a data collection module.
2. **Data Processing Module**: This module preprocesses the acquired data and subsequently, extracts the features from the processed data. Table 1 presents some of the common data preprocessing techniques to convert raw sensory data into an understandable format.
3. **Features Processing Module**: In this module, the extracted features are fused and selected for the generation of a user biometric template. Biometric traits may be acquired separately or simultaneously and they are processed as per the fusion model used.
4. **Database Module**: Database module stores the users' biometric template generated during the enrollment process.
5. **Classification Module**: Classification module compares the input query and stored biometric template of an individual to accept or reject the claimed identity.

Ten et al. [32] stated that access to critical CPS should be validated through the biometric trait of an individual. Furthermore, Ross et al. [30] elaborated that smart cities have a nexus of interconnected IoT devices that require human-machine and machine-machine interaction. Human-machine interaction can be secured via biometric verification to provide personalized services as well as to ensure residents' safety.

3 Problem Description

This section analyzes the potential threats to CPS facilitating smart financial solutions, smart transportation, and smart buildings. The main purpose of IAM schemes is verification, authorization, administration of identities, and audit. Thus, IAM

Table 1 Data preprocessing techniques

#	Technique	Description
1.	Data cleaning	Fill in missing values, smooth noisy data, identify or remove outlier, and resolve inconsistencies
2.	Data integration	Integration of data from multiple sensors
3.	Data transformation	Normalization and aggregation of sensor data
4.	Data reduction	Obtains reduced representation in volume but produces the same or similar analytical results
5.	Data discretization	A part of data reduction but with particular importance, especially for numerical data

schemes should ensure an appropriate level of security for these strategic solutions while simultaneously keeping schemes usable for end-users and administrators.

3.1 Smart Financial Solutions

A cashless environment for smart cities as an alternative to paper money can bring manyfold benefits to residents. It can provide hassle-free money transfers, bring down crimes like mugging, and even can restrict the spread of viruses in pandemic-like situations. Thus, smartphone-based payment apps like *Apple Pay Cash*, *Alipay*, *Google Pay*, *PayPal*, and *Samsung Pay* can be deemed essential for providing smart financial solutions.

Considering physical attacks, where (i) the adversary accidentally finds an unlocked smartphone, (ii) the adversary is a friend or colleague (who possibly knows the user's PIN/Passwords), and (iii) the adversary records users while they interact with their smart devices. Eventually, the adversary exploits the weaknesses of PIN/password-based verification schemes to gain access to users' smart payment apps.

Prior studies [29, 31] also demonstrated that the aforementioned scenarios are quite apparent, as users use their smart devices at commons places like offices, homes, meeting rooms, or streets, which may give opportunities to adversaries to target their smart devices, easily. As a consequence, smartphone users can be a victim of monetary fraud, identity thefts, or similar unfavorable incidents.

3.2 Smart Transportation

With the goals set for minimum congestion, accident-free travel, and safety of residents, smart transport is another important aspect towards the realization of smart cities. Unarguably, on-demand ride and ride-sharing services have revolutionized the point-to-point transportation market. Customers can book these rides services on short notice with 24×7 availability in all major cities across the world.

However, the reliability of drivers has emerged as a critical problem, and as a consequence, issues related to riders' safety and security have started surfacing. News related to fake drivers and assaults by dishonest drivers is a severe safety and security risk for the riders [35]. Further, being a lucrative business and easy to start, on-demand rides and ride-sharing services are attracting people also with unclean police records to become driver-partners using false identities [4]. Eventually, there can be two types of malicious users: the first type of adversary can impersonate a driver-partner by imitating a legitimate driver. The second type of attacker colludes with a legitimate driver-partner sharing the same registration to provide rides on behalf of the legitimate driver.

3.3 Smart Buildings

In today's rapidly evolving smart cities, frictionless and smooth interactions for smart buildings are emerging as critical a requirement. Such requirements need to coexist with mandatory properties like physical security. Consequently, smart buildings require some form of physical access control (i.e., locks, doors, barriers, etc.) that must be both reliable and usable for the users [22]. Any unauthorized access to smart buildings maintaining public utilities, e.g., healthcare, electricity, water, or gas, can disrupt the day-to-day life in a smart city.

4 IAM scheme for Smart Financial Solutions

We design a bimodal behavioral biometric-based one-shot-cum-continuous user verification scheme that authenticates users based on *how* they enter the text instead of *what* they enter, thus strengthening `username/password`-based schemes used in smartphone-based payment apps without incurring additional cost to the smart financial solution providers.

HOLD & TAP [7] strengthens the widely used PIN/password-based verification technology by giving flexibility to users to enter any random 8-digit alphanumeric text and authenticates users based on their invisible tap-timings and hand-movements, instead of pre-configured PIN or Passwords. Moreover, the entire user session is *continuously* safeguarded by assessing risk. Thus, HOLD & TAP, not only authenticates users during the application sign-in process but also, throughout the entire active user session.

Figure 2 illustrates the framework of our one-shot-cum-continuous verification scheme explaining how it addresses security and usability issues in existing `user/password`-based, and 2-factor verification schemes. The scheme enables users to enter any random $8 - digit$ alphanumeric text to access the application to

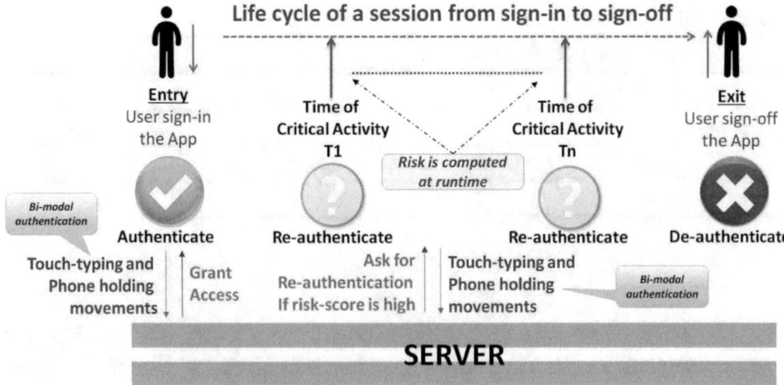

Fig. 2 HOLD & TAP verification scheme framework

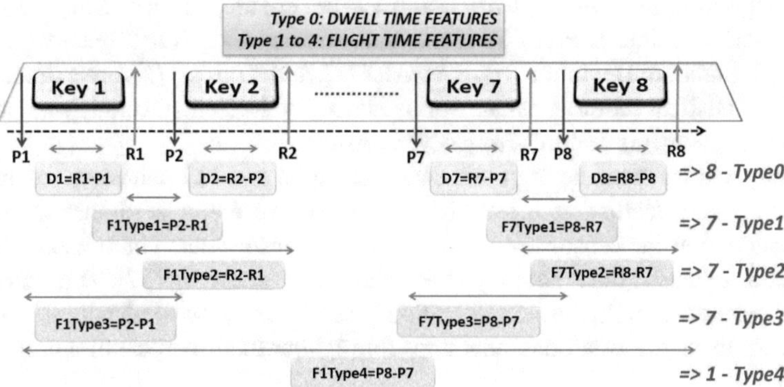

Fig. 3 Touch-typing features for 8-keys entry

enhance the usability of existing PIN/Password-based one-shot verification schemes. Consequently, the users' identities are verified based on timing differences between the entered keystrokes and their hand-movement in 3-dimensional space instead of just a binary comparison, to enhance security. After the successful sign-in, the scheme *continuously* monitors client attributes and computes the risk-score at the instant users initiate critical activities. Based on the risk-score, it permits users to perform that activity, otherwise, the scheme prompts for re-verification.

We extract 30 touch-typing features from the 8-digit *random-text* entry using touchscreen sensor. Figure 3 shows touch-typing features that consist of 8 *Type0* (timing difference between each key release and key press), 7 *Type1* (timing difference a key press and previous key release, 7 *Type2* (timing difference two successive keys release), 7 *Type3* (timing difference two successive keys press), and 1 *Type4* (timing difference between last and first key press).

A user's hand-movement signature is constructed from 4 raw data streams obtained from each of the seven motion sensors (*i.e., the accelerometer, the high-pass*

Table 2 Statistical features per sensor for a hand-movement behavior

No.	Hand-movement features			
1–4	μ_X	μ_Y	μ_Z	μ_M
5–8	σ_X	σ_Y	σ_Z	σ_M
9–12	s_X	s_Y	s_Z	s_M
13–16	k_X	k_Y	k_Z	k_M

sensor, the low-pass sensor, the orientation sensor, the gravity sensor, the gyroscope, and the magnetometer) with the delay set at SENSOR_DELAY_GAME [2]. The 4 raw data streams are X, Y and Z, and M. However, M (Magnitude) is computed mathematically using Equation $Value_M = \sqrt{(Value_x^2 + Value_y^2 + Value_z^2)}$. Where, $Value_M$ is the Magnitude and $Value_x$, $Value_y$ and $Value_z$ are the values of X, Y and Z value of a sensor, at a time t. From each raw data stream, 4 statistical features, namely Mean (μ), Standard Deviation (σ), Skewness (s), and Kurtosis (k), are extracted that gives 16 statistical features per sensor as shown in Table 2. Overall, we extracted 112 hand-movements features from seven sensors.

Finally, we concatenate 30 touch-stroke features and 112 hand-movements features to create a feature vector of size 142. Here, we prefer to choose the feature level fusion over the sensor level fusion because sensory data could have inconsistent and/or unusable data that may affect classifiers' accuracy [27]. Thus, HOLD & TAP [7] provides a multiclass classification based on users' two behavioral biometric modalities to secure smart payment apps throughout the life-cycle of a typical user session.

5 IAM Scheme for Smart Transportation

DRIVERAUTH [15] is a risk-based multi-modal verification scheme that exploits three biometric modalities, i.e., swipe gestures, *text-independent* voice and face, to make the on-demand ride and ride-sharing services secure and safer for their customers. DRIVERAUTH enrolls the drivers at the time of registration and verifies every time a new ride-assignment is given to them. Each smart transportation provider has its dedicated system and application for its driver-partners, however, the core functionalities are the same. Thus, DRIVERAUTH can easily be integrated into these systems and provide the required safety and security to customers.

Figure 4 explains the framework of DRIVERAUTH. A person intended to work as a driver-partner is registered to the system at the time of $Entry$, i.e., the person's biometric traits are acquired and added to the database for a reliable $1 - to - 1$ verification. According to ISO 9000:2015 [18], *risk* is the "effect of uncertainty on objectives". Here, a service provider accepting a driver-partner's ride request ($T_1 ... T_n$) can be considered as a critical activity. Therefore, verification of driver-partner to

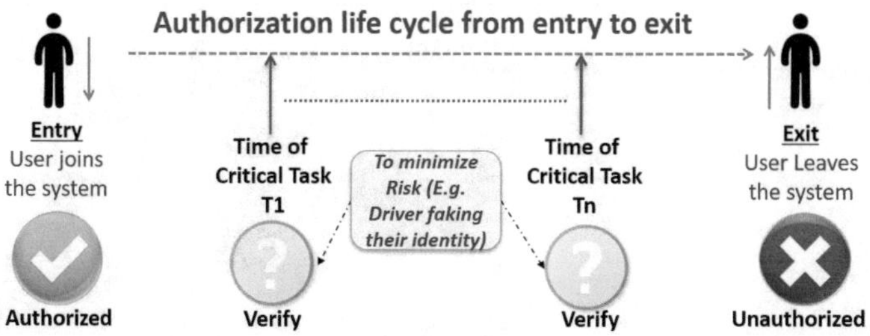

Fig. 4 DRIVERAUTH verification mechanism framework

Table 3 List of swipe features

No.	Swipe features			
1–4	Duration	Average event size	Event size down	Pressure down
5–8	Start X	Start Y	End X	End Y
9–12	Velocity X Min	Velocity X Max	Velocity X Average	Velocity X STD
13–16	Velocity X VAR	Velocity Y Min	Velocity Y Max	Velocity Y Average
17–20	Velocity Y STD	Velocity Y VAR	Acceleration X MIN	Acceleration X Max
21–24	Acceleration X AVG	Acceleration X STD	Acceleration X VAR	Acceleration Y MIN
25–28	Acceleration Y Max	Acceleration Y AVG	Acceleration Y STD	Acceleration Y VAR
29–32	Pressure min	Pressure max	Pressure AVG	Pressure STD
33	Pressure VAR	–	–	–

mitigate a potential risk must be performed. At the time of $Exit$, the driver-partner's biometric traits are deleted from the system and no more allowed to accept ride assignments.

The driver-partner swipe's gesture, i.e., a sequence of touch-events, is collected and encoded as an input sequence of finite length (n). Each sequence contains several attributes like time-stamp of the touch event (t_n), x-and y-coordinate of the touch point (x_n, y_n), pressure calculating how hard the finger was pressed on the screen (p_n), and size of touch area (s_n). We process the collected sequences and extract 33 features as listed in Table 3.

We acquire a two-second voiceprint of the driver-partner that contains 2 channels sampled at 44100 Hz with 16 bits per sample. The signal is first filtered using a band-pass filter to improve the signal-to-noise ratio. Then, we computed Mel Frequency Cepstral Coefficients (MFCC) from these filtered voice signals that are analogous to filters (vocal tract) in the source-filter model of speech. Figure 5 illustrates the

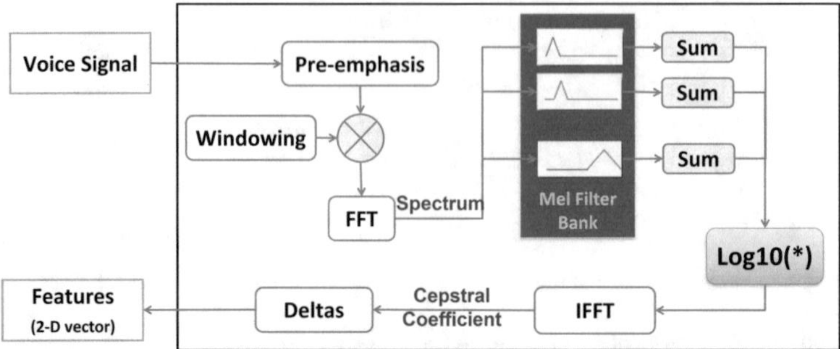

Fig. 5 Voice features: MFCC computation process

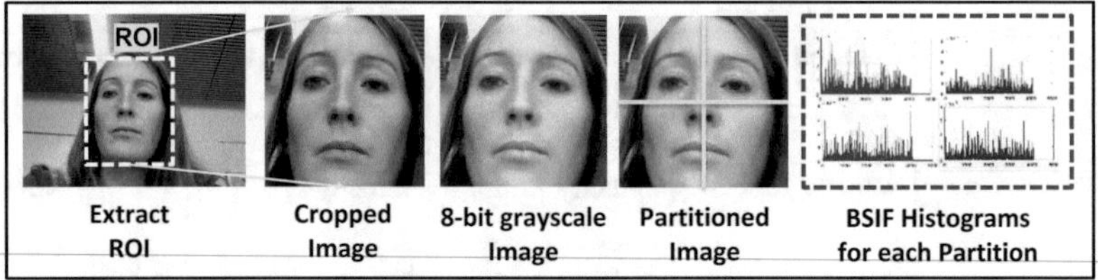

Fig. 6 BSIF features extraction process [26]

MFCCs computation process. The scaling of the frequency axis to the non-linear Mel scale (using triangular overlapping windows) is done after applying the Fourier transform on a window of the voice signal. After that, a Discrete Cosine Transform (DCT) is performed on the log of the power spectrum of each Mel band. The MFCCs are the amplitudes of the resulting spectrum, which is a $2 - D$ vector of size $13 \times variable\ length$ (the length of vector depends on the voice signal duration).

We compute 4 statistical features, namely mean, standard deviation, kurtosis, and skewness, from a 2-D MFCC vector. Thus, the total 8 statistical features each of size 1×13 are generated from each left and the right voice channel. Finally, these 8 vectors of size 1×13 are concatenated to form a single $1 - D$ feature vector of dimension 1×104.

Binarized Statistical Image Features (BSIF) extraction process for a face image is illustrated in Fig. 6. First, the driver-partner face image is cropped to extract the Region Of Interest (ROI). After that, each image is converted into an 8-bit grayscale format for obtaining statistical features using a BSIF filter bank [21]. Given an image patch X of size $l \times l$ pixels and a filter W of size $n \times n$ pixels, where n is less than l. The filter response s_i can be obtained using Equation $s_i = X[l, l] * W[n, n]$. We extract 256 features per image using a filter of size 3×3 with 8 bits word-length. BSIF filter applies learning, instead of manual tuning, to compute a statistically meaningful representation of an image.

DRIVERAUTH provides multi-model biometric-based user verification for proactive verification of driver-partners before allocating new ride assignments to them, thus, making smart transportation like on-demand ride and ride-sharing services safer and secure.

6 IAM Scheme for Smart Buildings

STEP & TURN [17] is a bimodal behavioral biometric-based verification system that utilizes two natural human actions, i.e., single footstep and hand-movement, to secure access to smart buildings. STEP & TURN exploits *single footstep* and *hand-movement* behavioral biometric for securing smart buildings. Both the footstep and hand-movement modalities do not require explicit users' cooperation and can be collected unobtrusively.

The door handle is fitted with three motion sensors that are accelerometer, gyroscope, and magnetometer. To model a user's hand-movement trajectory in 3-D space, `X`, `Y`, and `Z` data streams are acquired from the three sensors at the sampling of 50 Hz and `M` is derived mathematically. Subsequently, 6 independent univariate statistical features, namely, minimum (Min), maximum (Max), mean (μ), standard deviation (σ), kurtosis (k), and skewness (s) are computed from each data-stream using Eq. 1.

$$
\begin{aligned}
Min &= \min_{n=1}^{S} D_n & Max &= \max_{n=1}^{S} D_n \\
\mu &= \frac{1}{S}\sum_{n=1}^{S} D_n & \sigma &= \sqrt{\frac{\sum_{n=1}^{S}(D_i - \mu)}{S}} \\
k &= \frac{\frac{1}{S}\sum_{n=1}^{S}(D_n - \mu)^4}{\sigma^4} & s &= \frac{\frac{1}{S}\sum_{n=1}^{S}(D_n - \mu)^3}{\sigma^3}
\end{aligned}
\tag{1}
$$

where, S_i is the ith sample in a data-stream. N is the number of samples in a data-stream. Min and Max are the minimum and maximum values respectively, in a given data-stream. Mean (μ) is the average of all samples. Standard deviation (σ) is the square root of the variance. Kurtosis (k) measures the degree of peakedness of a data-stream that helps in detecting the outlier-proneness of the distribution. Skewness (s) measures the degree of asymmetry of a data-stream from its mean value. We extract 6 statistical features from each data-stream. As there are 4 data-streams per sensor so 24 (4×6) features are obtained per sensor. Thus, with 3 sensors, a total of 72 ($3 \times 4 \times 6$) statistical features are extracted. The final feature vector for hand-movement consists of 73 features in total including 72 statistical features and handle-movement action time

Footstep pressure-data acquisition system consists of 88 high-density piezoelectric sensors to acquire the footstep data. After that, each pressure amplitude array of size 88×2200 into 4-independent time-series arrays, namely, Spatial Average

(S_{ave}), Ground Reaction Force ($GRF_{cumulative}$), Upper (S_{upper}) and Lower (S_{lower}) Contours of size 1 × 2200 each [11], using Eq. 2.

$$S_{ave}[t] = \sum_{i=1}^{N} S_i[t] \qquad GRF_{cumulative}[t] = \sum_{t=1}^{T_{max}} S_{ave}[t]$$
$$S_{upper}[t] = \max_{i=1}^{N} S_i[t] \qquad S_{lower}[t] = \min_{i=1}^{N} S_i[t] \tag{2}$$

where, $S_i[t]$ is the differential pressure sample from the ith piezoelectric sensors at the time t. N is the total number of piezoelectric sensors, i.e., 88. Then, 6 statistical features are computed from each time-series array by using Eq. 1. In total, we get 48 statistical features (8 × 6) from 8 time-series arrays that were obtained from both left and right pressure amplitude arrays.

Ground Reaction Force (GRF_i) per sensor is computed by accumulating each sensor pressure amplitude from time T_1 to T_{max} by using 3.

$$GRF_i = \sum_{t=1}^{T_{max}} S_i[t] \tag{3}$$

where, $S_i[t]$ is the differential pressure sample from the ith piezoelectric sensors with i ranges from 1 to 88 and t ranges from 1 to 2200. In total, 176 features are obtained from both left and right pressure amplitude arrays.

Overall, STEP & TURN leverages 297 features extracted for hand-movement and footstep behavioral modalities to provide a multi-class classification solution for securing access to smart buildings.

7 Challenges and Limitations

In this section, we discuss challenges and limitations that can adversely affect CIA principles, i.e., confidentiality (*ensuring access to legitimate users only*), integrity (*guaranteeing modification by legitimate users*), and availability (*ensuring uninterrupted availability to legitimate users*) for designing biometric-based IAM schemes.

- Security analysis of biometric-based IAM schemes can be a challenging task, therefore, a thorough testing strategy must be developed for mitigation of vulnerability-detection, intra-class variance, and common attacks (e.g., malware, mimics, impersonation, spoofing, replay, statistical, algorithmic, and robotics attacks).
- Factors like aging, fatigue, illness, injury, mood, stress, or sleep deprivation, may impede the effectiveness of biometrics. These factors require in-depth investigation to support the development of biometric-based IAM schemes.
- Behavioral biometrics datasets must consider all demographics covering different age groups, cultural factors, and ethnicity to provide better objectivity. Moreover,

standards for behavioral biometrics and benchmarking of sensors must be developed and utilized.

- Quality control of the biometric template is a prerequisite before the enrollment or verification/identification phase. It can support the accuracy, stability, redundancy, and speed of IAM schemes to address problems arising from the environment, sensors, or the users themselves.
- Several privacy regulation laws such as General Data Protection Regulation (GDPR), the California Consumer Privacy Act (CCPA), and the Health Insurance Portability and Accountability Act (HIPAA) described biometric mandated an increase in responsibility and transparency for using and storing personal data [16]. Therefore, adequate measures like template protection and local template storage must be employed for complying with privacy laws.
- Ethical conduct in the use of behavioral biometrics is another important aspect that needs to be addressed carefully. For instance, acquiring and accumulating behavioral biometric data over time can lead to dynamic behavior profiling of an individual, providing insight into how the individual behaved in a certain context. This can be more problematic if behavioral biometric data is combined with soft biometrics, such as age, gender, height, weight, and ethnicity, to generate an individual's profile that can aggregate ethical risks.

8 Conclusions

Usable security emerged as a substantial requirement for a secure and safe smart city. We presented some next-generation biometric-based IAM schemes, namely, HOLD & TAP, DRIVERAUTH, and STEP & TURN for smart financial solutions, smart transportation, and smart buildings, respectively. HOLD & TAP can provide a pleasant and satisfying user experience by giving the flexibility to enter random alphanumeric text to access security-sensitive applications. Internally, users' invisible tap-timings and hand-movements are exploited to secure access to smart financial solutions. DRIVERAUTH can contribute to preventing unforeseen incidents to secure smart transportation by implementing risk-based multi-modal biometric-based. It proactively verifies driver-partners before new ride assignments are allocated to them. Lastly, STEP & TURN is a bimodal behavioral biometric-based verification system that can offer an easy access mechanism to secure smart buildings.

References

1. Aldawood H, Skinner G (2018) Educating and raising awareness on cyber security social engineering: a literature review. In: Proceedings of the IEEE international conference on teaching, assessment, and learning for engineering (TALE). IEEE, pp 62–68

2. Android: motion sensors. https://developer.android.com/guide/topics/sensors/sensors_motion. Accessed on 20 Feb 2022
3. Antonakakis, M.: Understanding the Mirai botnet. In: Proceedings of the 26th USENIX security symposium, pp 1093–1110 (2017)
4. BBC (2015) Uber driver background checks not good enough. http://www.bbc.com/news/technology-34002051. Accessed 20 Feb 2022. Online web resource
5. Binbeshr F, Kiah MM, Por LY, Zaidan AA (2021) A systematic review of pin-entry methods resistant to shoulder-surfing attacks. Comput Secur 101:102116
6. Braz C, Seffah A, Naqvi B (2018) Integrating a usable security protocol into user authentication services design process
7. Buriro A, Gupta S, Yautsiukhin A, Crispo B (2021) Risk-driven behavioral biometric-based one-shot-cum-continuous user authentication scheme. J Signal Process Syst
8. Choi H, Kwon H, Hur J (2015) A secure OTP algorithm using a smartphone application. In: Proceedings of the 7th international conference on ubiquitous and future networks. IEEE, pp 476–481
9. Dasgupta D, Roy A, Nag A et al (2017) Advances in user authentication
10. Dilraj M, Nimmy K, Sankaran S (2019) Towards behavioral profiling based anomaly detection for smart homes. In: Proceedings of the TENCON 2019-2019 IEEE region 10 conference (TENCON). IEEE, pp 1258–1263
11. Edwards M, Xie X (2014) Footstep pressure signal analysis for human identification. In: Proceedings of the 7th international conference on biomedical engineering and informatics. IEEE, pp 307–312
12. El-Hajj M, Fadlallah A, Chamoun M, Serhrouchni A (2019) A survey of internet of things (IoT) authentication schemes. Sensors 19(5):1141
13. Gamundani AM, Phillips A, Muyingi HN (2018) An overview of potential authentication threats and attacks on internet of things (IoT): a focus on smart home applications. In: Proceedings of the IEEE international conference on internet of things (iThings) and IEEE green computing and communications (GreenCom) and IEEE cyber, physical and social computing (CPSCom) and IEEE smart data (SmartData). IEEE, pp 50–57
14. Gupta S (2020) Next-generation user authentication schemes for IoT applications. PhD thesis, DISI, Univeristy of Trento, Italy
15. Gupta S, Buriro A, Crispo B (2019) Driverauth: a risk-based multi-modal biometric-based driver authentication scheme for ride-sharing platforms. Comput Secur 83:122–139
16. Gupta S, Camilli M, Papaioannou M (2022) Provenance navigator: towards more usable privacy & data management strategies for smart apps. In: Proceedings of the 11th international workshop on socio-technical aspects in security, affiliated with the 26th European symposium on research in computer security (ESORICS 2021). Springer, pp 1–17
17. Gupta S, Kacimi M, Crispo B (2022) Step & turn—a novel bimodal behavioral biometric-based user verification scheme for physical access control. Comput Secur
18. ISO9000:2015 (2015) Quality management systems—fundamentals and vocabulary. https://www.iso.org/obp/ui/iso:std:iso:9000:ed-4:v1:en. Accessed on 20 Feb 2022. Online web resource
19. ISO/IEC24741:2018(en) (2018) Information technology—biometrics—overview and application. https://www.iso.org/obp/ui/iso:std:iso-iec:tr:24741:ed-2:v1:en
20. Jain AK, Deb D, Engelsma JJ (2021) Biometrics: trust, but verify. IEEE Trans Biom Behav Identity Sci
21. Kannala J, Rahtu E (2012) Bsif: binarized statistical image features. In: Proceedings of the 21st international conference on pattern recognition (ICPR). IEEE, pp 1363–1366
22. Krašovec A, Pellarini D, Geneiatakis D, Baldini G, Pejović V (2020) Not quite yourself today: behaviour-based continuous authentication in IoT environments. Proc ACM Interact Mob Wearable Ubiquitous Technol 4(4):1–29
23. Li W, Wang P (2019) Two-factor authentication in industrial internet-of-things: attacks, evaluation and new construction. Futur Gener Comput Syst 101:694–708

24. Liang X, Kim Y (2021) A survey on security attacks and solutions in the IoT network. In: Proceedings of the 11th annual computing and communication workshop and conference (CCWC). IEEE, pp 0853–0859
25. Ling Z, Liu K, Xu Y, Jin Y, Fu X (2017) An end-to-end view of IoT security and privacy. In: Proceedings of the GLOBECOM 2017—2017 IEEE global communications conference, pp 1–7
26. McCool C, Marcel S, Hadid A, Pietikäinen M, Matejka P, Cernockỳ J, Poh N, Kittler J, Larcher A, Levy C et al (2012) Bi-modal person recognition on a mobile phone: using mobile phone data. In: Proceedings of international conference on multimedia and expo workshops (ICMEW). IEEE, pp 635–640
27. Pires I, Garcia N, Pombo N, Flórez-Revuelta F (2016) From data acquisition to data fusion: a comprehensive review and a roadmap for the identification of activities of daily living using mobile devices. Sensors 16(2):184
28. Ponnusamy V, Regunathan ND, Kumar P, Annur R, Rafique K (2020) A review of attacks and countermeasures in internet of things and cyber physical systems. Industrial internet of things and cyber-physical systems: transforming the conventional to digital, pp 1–24
29. Project OMS (2020) Owasp mobile security project. https://owasp.org/www-project-mobile-security/. Accessed 20 Feb 2022. Online web resource
30. Ross A, Banerjee S, Chowdhury A (2020) Security in smart cities: a brief review of digital forensic schemes for biometric data. Pattern Recognit Lett 138:346–354
31. Shila DM, Srivastava K (2018) Castra: seamless and unobtrusive authentication of users to diverse mobile services. IEEE Internet Things J 5(5):4042–4057
32. Ten CW, Manimaran G, Liu CC (2010) Cybersecurity for critical infrastructures: attack and defense modeling. IEEE Trans Syst Man Cybern Part A Syst Hum 40(4):853–865
33. Van Oorschot PC (2021) User authentication-passwords, biometrics and alternatives. In: Proceedings of the computer security and the internet. Springer, Cham, pp 55–90
34. Verizon. Data breach investigations report. https://enterprise.verizon.com/content/verizonenterprise/us/en/index/resources/reports/2021-dbir-executive-brief.pdf. Accessed on 20 Feb 2022. Online web resource
35. Whosdrivingyou (2018) Reported list of incidents involving uber and lyft. http://www.whosdrivingyou.org/rideshare-incidents. Accessed on 20 Feb 2022. Online web resource
36. Zhang K, Ni J, Yang K, Liang X, Ren J, Shen XS (2017) Security and privacy in smart city applications: challenges and solutions. IEEE Commun Mag 55(1):122–129
37. Zimmermann V, Gerber N (2020) The password is dead, long live the password—a laboratory study on user perceptions of authentication schemes. Int J Hum Comput Stud 133:26–44

Collaborative Security Patterns for Automotive Electrical/Electronic Architectures

Florian Fenzl, Christian Plappert, Roland Rieke, Daniel Zelle, Gianpiero Costantino, Marco De Vincenzi, and Ilaria Matteucci

Abstract In this chapter, we describe several security design patterns that collaboratively consider various cybersecurity aspects with the aim to ensure compliance with cybersecurity requirements for a certified cybersecurity and software update management system imposed by the recent United Nations regulations. Automated driving requires increasing networking of vehicles, which in turn expands their attack surface. The security design patterns enable the detection of anomalies in the firmware at boot, ensure secure communication in the vehicle and detect anomalies in in-vehicle communications, prevent unauthorized electronic control units from successfully transmitting messages, provide a way to transmit and aggregate security-related events within a vehicle network, and report to entities external to the vehicle. Using the example of a future high-level automotive Electrical/Electronic architecture, we also describe how these security design patterns can be used to become aware of the current attack situation and to react to it.

F. Fenzl · C. Plappert · R. Rieke · D. Zelle
Fraunhofer SIT, Darmstadt, Germany
e-mail: florian.fenzl@sit.fraunhofer.de

C. Plappert
e-mail: christian.plappert@sit.fraunhofer.de

R. Rieke
e-mail: roland.rieke@sit.fraunhofer.de

D. Zelle
e-mail: daniel.zelle@sit.fraunhofer.de

G. Costantino (✉) · M. De Vincenzi · I. Matteucci
Istituto di Informatica e Telematica, Consiglio Nazionale delle Ricerche, Rome, Italy
e-mail: gianpiero.costantino@iit.cnr.it

M. De Vincenzi
e-mail: marco.devincenzi@iit.cnr.it

I. Matteucci
e-mail: ilaria.matteucci@iit.cnr.it

T. Dimitrakos et al. (eds.), *Collaborative Approaches for Cyber Security in Cyber-Physical Systems*, Advanced Sciences and Technologies for Security Applications,
https://doi.org/10.1007/978-3-031-16088-2_4

1 Introduction

Automated driving requires increasing networking of vehicles, which in turn expands their attack surface. To counteract this threat, the United Nations (UN) regulations on cybersecurity management [35] and software update management [36] introduce new requirements for the consideration of cybersecurity aspects in vehicle type approvals. Specific requirements for vehicle types include first identifying the critical elements of the vehicle type and conducting a comprehensive risk assessment taking into account the individual elements of the vehicle type, their interactions and interactions with external systems. In order to protect the vehicle type from identified risks, the manufacturer must implement measures to (a) detect and prevent cyberattacks on vehicles of the vehicle type; (b) to ensure monitoring with regard to the detection of threats, vulnerabilities and cyberattacks relevant to the vehicle type; (c) provide data forensics to enable analysis of attempted or successful cyberattacks.

Based on our own preliminary work in [24], in this chapter, we propose several security patterns that can be used when designing a new automotive Electrical/Electronic (E/E) architecture for a vehicle type that meets the above UN requirements. The overall contributions of our work include multiple security design patterns and their collaborative interaction, with the aim of targeting critical steps in attack chains and mitigating their consequences. Using the example of a future high-level E/E vehicle architecture, we also describe how these security design patterns can be used to become aware of the current attack situation and to react to it.

This chapter is structured as follows. In Sect. 2 we present our approach and detail of our patterns, which consist of a security event reporting pattern (Sect. 2.1), a host-based integrity verification pattern (Sect. 2.2), a key management pattern (Sect. 2.3), a network-based intrusion detection pattern (Sect. 2.4), a challenge-based intrusion prevention pattern (Sect. 2.5), a pattern for securing in-vehicle communications (Sect. 2.6), a pattern for secure service discovery (Sect. 2.7), and a pattern for secure in-vehicle feature activation (Sect. 2.8). In Sect. 3 we map the capabilities of the patterns to structural cyber resiliency design principles, and in Sect. 4 we show an example for collaborative application of the proposed patterns for attack detection and mitigation in an automotive E/E reference architecture. We conclude this chapter by a summary in Sect. 5 and provide some bibliographic references in Sect. 6.

2 Patterns for Designing Collaborative Concepts to Secure Automotive Architectures

We propose the following patterns for the design of collaborative concepts for robust attack detection and mitigation for future automotive E/E architectures.

Security Event Reporting (SER): This pattern provides a way to aggregate security relevant events, such as software anomalies or unexpected network traffic, within a vehicle network and report it to an outside Security Operation Center (SOC).

Host-based Integrity Verification (HbIV): This pattern provides a mechanism to detect anomalies (unauthorized modifications) in Electronic Control Unit (ECU) software and firmware at boot time. The pattern has two variants, using the Trusted Platform Module (TPM) as hardware trust anchor or the Device Identifier Composition Engine (DICE) as lightweight alternative.

Key Management (KM): This pattern provides a mechanism to manage the exchange and storage of keys used by ECUs to secure intra-vehicle messages.

Network-based Intrusion Detection (NbID): This pattern provides a mechanism to detect anomalies in automotive network communication by using metadata and extracted payload features from observed ECU messages.

Challenge-based Intrusion Prevention (CbIP): This pattern provides a mechanism to prevent anomalies in the automotive network communication by using a challenge based on the knowledge of the payload to authenticate an ECU.

Secure CAN Communication (SCC): This pattern provides a mechanism to secure the in-vehicle communication. It guarantees confidentiality, integrity, and authentication.

Secure Service Discovery (SSD): This pattern provides a secure mechanism to propagate and find service offer in an in-vehicle network. It guarantees integrity and authentication of the offered services.

Secure Feature Activation (SFA): This pattern provides a secure mechanism for authorized users to activate their (purchased) features in the vehicle. It guarantees integrity and authenticity of the protocols and components.

Below we describe each pattern in detail using the automotive security template defined in [7] with the following elements: *Pattern Name*, *Intent* describing the underlying security problem addressed by the pattern, *Motivation*, *Properties* in terms of STRIDE [30], *Applicability*, *Structure* given by a Unified Modeling Language (UML) class diagram, *Behavior* defined by a UML sequence diagram, *Constraints*, *Consequences* expressed by security properties, performance, cost, manageability, usability, *Known Uses*, and *Related Patterns*.

The security patterns described in this chapter are striving for compatibility with the Automotive Open Systems Architecture (AUTOSAR) standard. AUTOSAR, a global development partnership of vehicle manufacturers and related companies, specifies an in-vehicle Intrusion Detection System (IDS) protocol [3] for the transmission of qualified security events from a IDS manager instance to an IDS reporter instance and a IDS manager for the adaptive platform [2], a specification of Secure Onboard communication (SecOC) [4], as well as a Key Manager [5].

2.1 Security Event Reporting (SER)

Intent. This pattern provides a way to aggregate security relevant events, such as software anomalies or unexpected network traffic, within the vehicle network at a central point and to report it to an outside SOC. This enables experts to make collaborative decisions based on distributed analytical knowledge.

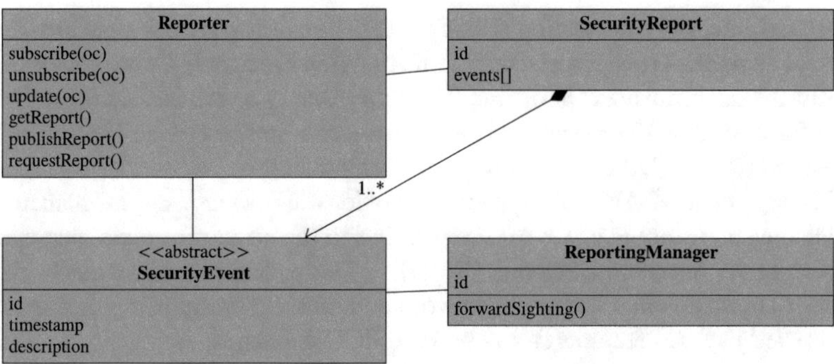

Fig. 1 Structure of the SER pattern

Motivation. A solely autonomous and isolated anomaly detection software is in most cases not feasible within an automotive network. This is especially the case when mitigation operations are set to occur as a response to critical security events. An external SOC is required to oversee mitigation actions and adapt classifiers to new situations. To achieve this, security events from multiple vehicles have to be accumulated, analyzed and verified by human operators in addition to automated systems.

Recorded data may also be used at a later stage in forensic analysis in case some incident occurs regarding the vehicle. Logged anomaly events could give directions on what happened within the vehicle and therefore help operators reconstruct the chain of events leading to the incident.

Properties. The Security Event Reporting (SER) pattern can influence the integrity and non-repudiation properties.

Applicability. The SER pattern is applicable to detection of anomalous behavior, as well as the prevention and mitigation of potential attacks.

Structure. The structure of the Reporter component is depicted in Fig. 1 and should be integrated on the primary component responsible for outside connection, e.g., the Telematic Control Unit (TCU). A set of distributed security components capable of reporting Security Events, such as observed anomalies, in a pre-defined structure are deployed throughout the vehicular network. A Security Event is a uniquely identifiable data object on a singular security relevant observation with a meaningful description and an exact time of occurrence. These Security Events can then be accumulated into uniquely identifiable collaborative Security Reports, which can then be used to forward compressed information to a remote observer, e.g., SOC or Original Equipment Manufacturer (OEM) backend.

Behavior. The specific behavior is described by Fig. 2. Whenever the Reporter component receives a sighting from any Reporting Manager within the network, either as a direct response or as a timed event, a report of accumulated sightings is created. When a new report is created, all subscribed SOCs are notified and may request the new sighting report. A SOC can also actively request the latest report from the component.

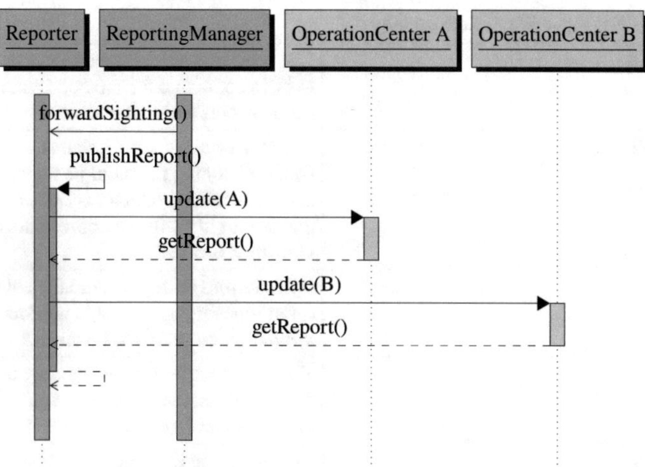

Fig. 2 Behavior of the SER pattern

Constraints. The Reporter component provides intruders with a potential centralized point of attack. This problem may be addressed by deploying multiple redundant instances of the Reporter within the vehicle, though this would increase network traffic and require an approach to synchronize the instances. For the Security Events send from a Reporting Manager (RM) to the Reporter, additional verification steps may be required to ensure integrity.

Consequences. See Table 1.

Known Uses. In their work, Metzker et al. presented their ideas, concepts and software architecture proposals concerning IDS and intrusion reporting in [22], while AUTOSAR provides a specification of an in-vehicle IDS protocol in [3] with a use case which covers event reporting similar to the proposed pattern.

Related Patterns. This pattern relates to all intrusion detection patterns, such as the Host-based Integrity Verification (HbIV), NbID, and CbIP pattern by providing a way to accumulate the data of all sensors. The Signature-based IDS pattern proposed by Cheng et al. [7] can also be integrated similarly with minimal modification.

2.2 *Host-based Integrity Verification (HbIV)*

Intent. With the HbIV pattern, unauthorized changes in ECU software and firmware at boot time can be detected. This is done by creating a trusted measurement chain starting from an initial Root of Trust for Measurement (ROTM) (e.g., in the boot loader), where the software component, which will be executed next, is measured (by computing a hash value of the software component) before control is transferred to it. The resulting measurements cover the entire software state during the boot process.

Table 1 Consequences of the SER pattern

Accountability	The reporting component allows for the system to be accountable for the non-repudiation of traffic events inside the vehicle
Confidentiality	Security event reporting shares vehicle internal traffic including potential personal data with external entities, which may prevent confidentiality without appropriate data security measures
Integrity	By reporting detected anomalies, the security event reporting may enable appropriate mitigation measures to be taken
Availability	The security event reporting may prevent availability in the vehicle network depending on the number of security events raised
Performance	Depending on the urgency of detected anomaly sightings there is an additional load on the vehicle network, while the reporter itself requires low computational power
Cost	The system itself requires low computational performance and can be integrated into an existing network that provides the required outside connections and connection to reporting components
Manageability	Allows for possible mitigation actions to be more manageable and better directed
Usability	Not addressed

This software state can then be reported to remote entities like OEM backends, Vehicle to Everything (V2X) participants, or SOCs.

We propose both a TPM-based [34] and a DICE-based [32] instantiation for the HbIV pattern. Both instantiations provide different security guarantees that account for the heterogeneous ECU capabilities of the vehicle. The TPM-based instantiation utilizes the TPM as dedicated security chip as Root of Trust for Storage (ROTS) for storage and as Root of Trust for Reporting (ROTR) for reporting of the measurements to remote entities. The DICE-based instantiation is a lightweight alternative using software mechanisms bootstrapped from a secure ROTM. Since TPMs are already started to be used in vehicles [16] and DICE has minimum silicon requirements and is inexpensive to integrate, both options are suitable for automotive applications. We explain the differences of both instantiations throughout this section.

Motivation. With the increasing autonomy of the vehicles, trust in the integrity and authenticity of vehicle systems is important since they more and more effect also the safety of all road users.

Properties. The HbIV pattern allows proving the integrity and authenticity of vehicle systems, e.g., ECUs, to remote verifiers.

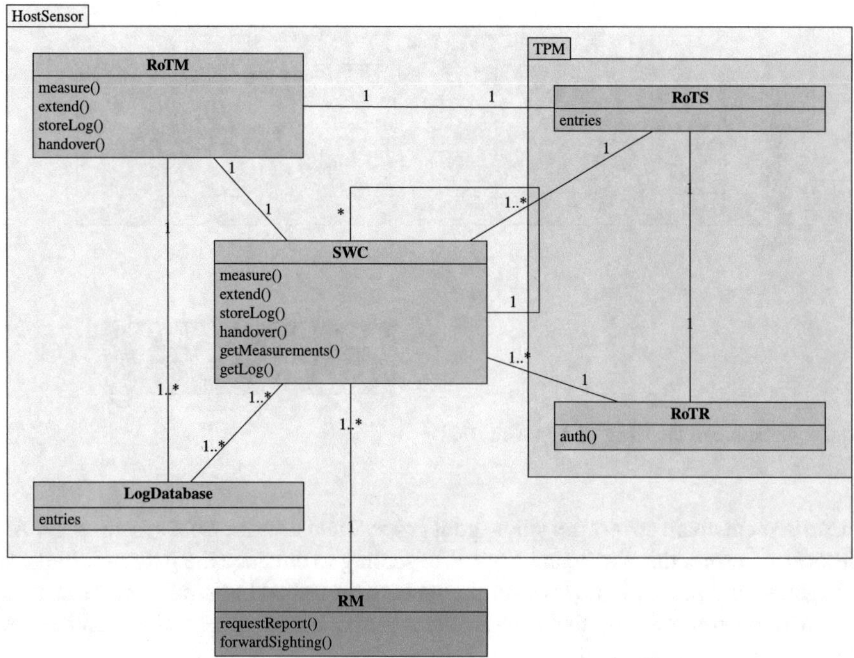

Fig. 3 Structure of the HbIV pattern with TPM

Applicability. In its basic form, the HbIV pattern is appropriate to detect and report attacks. An example is, to secure Over-the-Air (OTA) update mechanisms according to UNECE 156, by verifying that a correct software version was installed. However, the underlying trusted computing technology (TPM and DICE) can be easily utilized to enhance the pattern for attack mitigation and prevention, e.g., with sealing concepts where decryption or authentication keys can be unlocked only in an unmodified software state of the ECU.

Structure. The structure of the pattern is shown in Fig. 3 for the TPM-based and in Fig. 4 for the DICE-based instantiation of the pattern HbIV. As seen, both variants roughly consist of two main components. These are the host sensor and the RM. The latter is part of the Security Event Reporting (SER) pattern (cf. Sect. 2.1) and connects both patterns. Thus, the actual HbIV pattern is the host sensor that is associated with a specific ECU. In the TPM variant, the sensor consists of TPM, that provides the ROTM and ROTS, ROTR, log database, and the Software Component (SWC). In the DICE variant, the sensor is simpler and has only ROTM and SWC as components.

Behavior. Typically, the Central Processing Unit (CPU) acts as ROTM that executes the initial measurement of the first SWC (SWC1) before passing control to SWC1. SWC1 continues the measurement process with the next component (SWC2). This process is continued for all following software components, so that the final

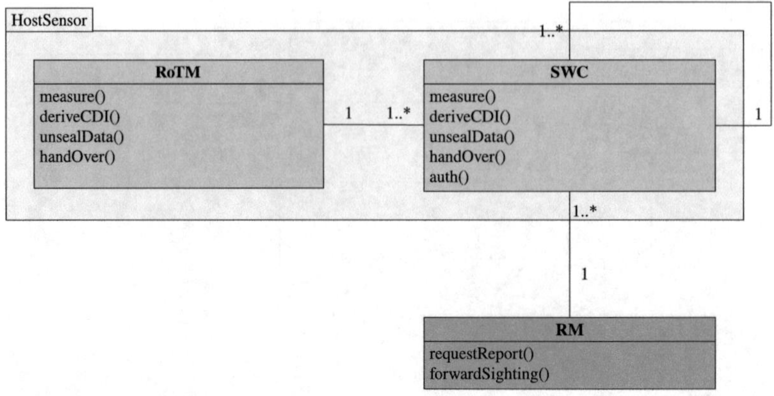

Fig. 4 Structure of the HbIV pattern with DICE

measurement chain covers the whole boot process and thus the boot system state. As described before, this state can be used for sealing to enhance the pattern for attack mitigation and prevention. However, in the here described basic pattern, the state is reported to a remote party that is then able to verify the integrity and authenticity of the system (remote attestation).

Details about the behavior of the TPM instantiation of the HbIV pattern are shown in Fig. 5. Here, the measurements of the SWCs are extended as hash values into the TPM's ROTS. Enriched with some metadata, the hash values are also stored in the log database. If an external party initiates a remote attestation, the measurements are authenticated by the ROTR, e.g., via Message Authentication Code (MAC) or digital signature, before exiting the ROTS.

The DICE instantiation is shown in Fig. 6. Its behavior is a bit different and also simpler. The integrity measurements of the various SWCs are not stored in a ROTS but used as inputs of the key derivation functions of the different SWCs to derive software state-dependent keys, so-called CDIs. For the initial Compound Device Identifier (CDI) is derived from a Unique Device Secret (UDS). The integrity and authenticity of the system can then be proven if the correct keys are derived, e.g., in a challenge-response-based remote attestation scheme initiated by a remote verifier.

In both instantiations, the authenticated events (answered challenges or measurement logs) can be sent to an external verifier for remote attestation. Thus, either actively or upon request, the RM collects the events from all sensors and sends them back via Reporter. The remote verifier can then use the measurements to decide if the requested component is trustworthy.

Constraints. The hash operations used in both instantiations during system boot are lightweight regarding processing power and memory consumption. Depending on the granularity and complexity of the measured component, the creation of the log database in the TPM instantiation is more demanding. But this could also be outsourced to a backend system that preemptively creates the expected values.

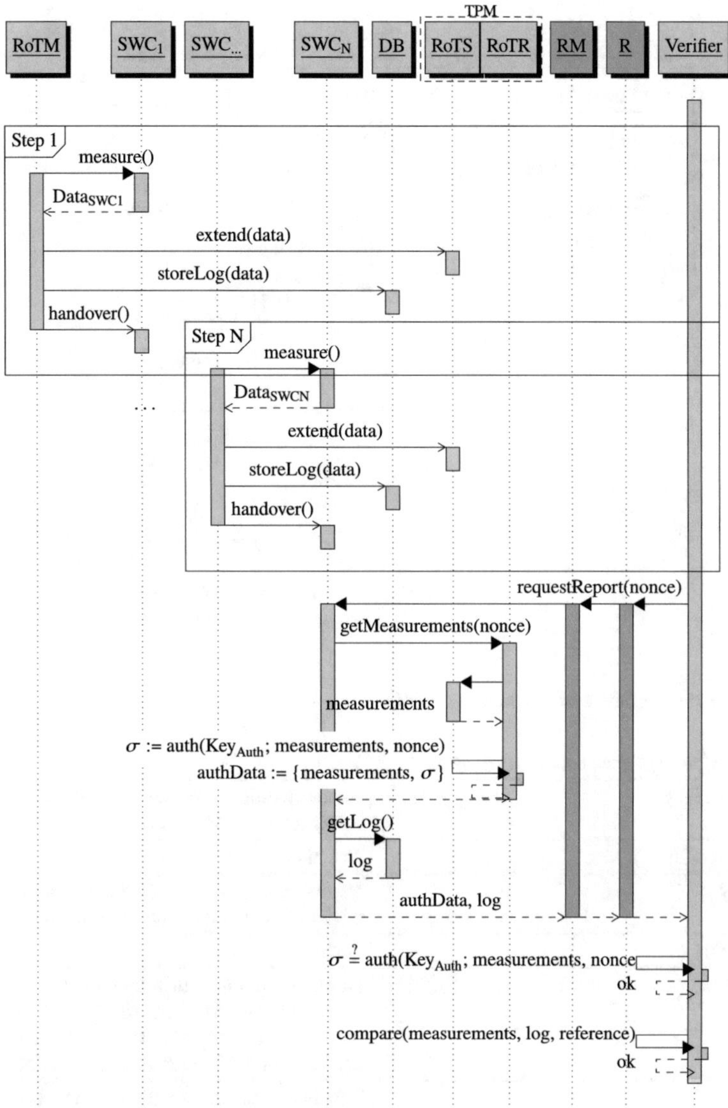

Fig. 5 Behavior of the HbIV pattern with TPM

In both instantiations, the measurements are authenticated during the reporting phase. Depending on the cryptographic schemes and reporting frequency, this may result in some additional performance overhead. The work in [13] presents an approach to make the reporting process more efficient by authenticating measurements in advance.

Consequences. Table 2 describes the consequences.

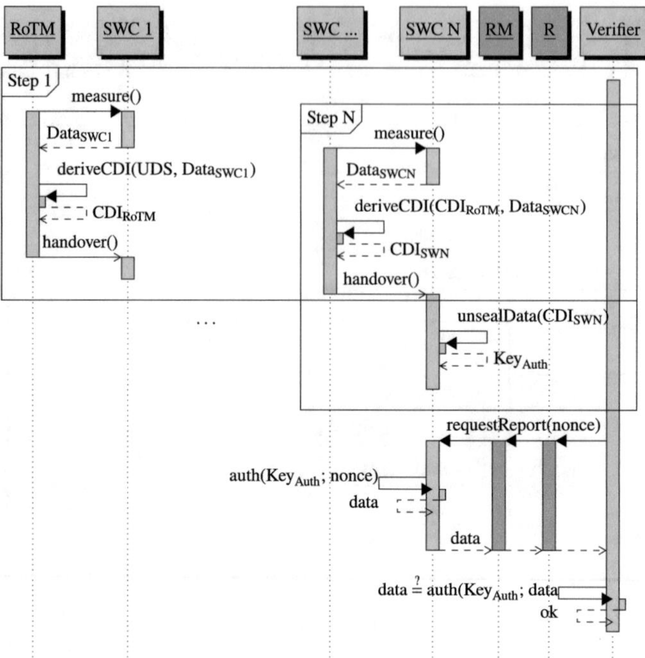

Fig. 6 Behavior of the HbIV pattern with DICE

Table 2 Consequences for the HbIV pattern

Accountability	Accountability is improved since the TPM/DICE establishes trust in the system
Confidentiality	Not directly addressed
Integrity	Remote parties can verify the trustworthiness of the vehicle, which improves the integrity
Availability	Not directly addressed
Performance	During boot time lightweight symmetric cryptographic operations (hashes) and during runtime more resource-intensive (a-)symmetric cryptography (signature/mac generation)
Cost	The TPM instantiation has additional hardware costs
Manageability	Standardized Application Programming Interface (API) and extensive software ecosystem facilitate management
Usability	The TPM chip introduce some overhead regarding performance and cost to the system

Known Uses. Parts of the described attestation schemes for the HbIV instantiations are standardized by Trusted Computing Group (TCG) and Internet Engineering Task Force (IETF), e.g., in [27, 33]. Applications for the automotive domain are described in [25] for feature activation and in [14] for secure updates.

Related Patterns. A directly related pattern is the *Security Event Reporting* pattern that is used to collect and send the measurements to remote parties. The patterns *Tamper Resistance* and *Third-party Validation* of [7] partially address similar security goals.

2.3 Key Management (KM)

Intent. This pattern provides strategies to manage keys in the light of the constraints and requirements that the automotive domain imposes.

Motivation. Modern vehicles' functionalities are managed and regulated via Electronic Control Units (ECUs) that interact to one another by using communication protocol such as, CANs and CAN-FDs. However, the ECUs bus protocol is not secure by design: it lacks of authentication, confidentiality and integrity. Existing protocols that implement secure communication among the various ECUs use secret key to secure exchanged messages. However, there is no phase in which such key is shared, thus making it a preliminary part of the actual communication. The problem of session key sharing is at the heart of the implementation of secure communication.

Properties. Over the in-vehicle communications, it is possible to apply different communication protocols that have to assure authentication, message integrity, and confidentiality.

Applicability. The Key Management (KM) pattern can assure the previous properties, managing keys that should be used by ECUs to generate and send secure messages, or when the secure message is received to get its content. Using encryption mechanisms may prevent threats like spoofing or replay.

Structure. Figure 7 shows the structure of the KM pattern. In an in-vehicle network, every ECUs can generate a secure frame to send to another ECUs. Besides, ECUs can create *vanilla* protocol frames to guarantee backward compatibility.

In the KM pattern, we have three main actors: the Key Manager (KM) that manages the keys, the key master that generate the key materials, and the Crypto Service Manager (CSM) that provides the security keys to the ECUs involved in the communication that must share the same keys to communicate securely.

Behavior.

Figure 8 shows the behavior of the KM pattern. The Key Management module is used to initialize, update and maintain cryptographic keys for the involved ECUs. The Key Management module will use a key master to provision the key material that will be used within the secure communication among ECUs. One use case is the provision of keys for the SCC pattern that need to be distributed to the involved ECUs. These keys should be provided to a service manager component, such as a Crypto Service Manager (CSM), that will use the respective keys to secure messages.

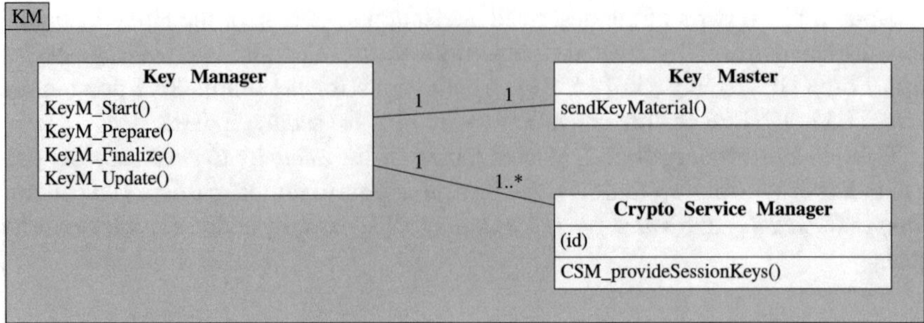

Fig. 7 Structure of the KM pattern

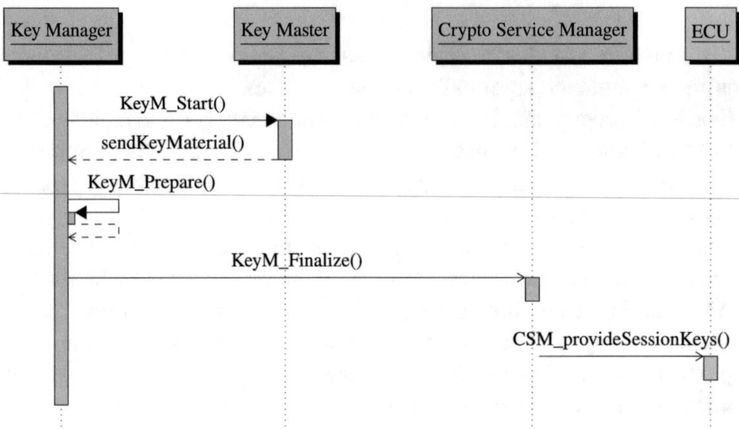

Fig. 8 Behavior of the KM pattern

This is an overall task in a vehicle and affects several ECUs in the same way. The key master can either be located directly in the vehicle to coordinate the key generation internally, e.g., as a particular ECUs. It is also possible to use a backend system in the cloud that generates the key material and provides the necessary data securely to the ECUs. The CSM is a service that provides cryptography functionality. The main task of the CSM is to use the locally stored keys and to provide the needed interface to the SCC pattern to generate secure messages. The interfaces are represented by those libraries that permits the invocation of cryptographic primitives. Such primitives can be both software or hardware.

Constraints. To work, the KM pattern can be costly in terms of overall system performance. In an automotive system where constraints exist on real-time processing and resources, the algorithms to create, exchange, and manage the key can represent a real critical aspect.

Table 3 Consequences of the KM pattern

Accountability	KM is accountable since the system is able to more accurately authenticate actors
Confidentiality	Frames that may be accessible from a possible attacker are more secured
Integrity	Frames integrity can be better protected one encrypted using keys
Availability	KM does not prevent availability, even if it may add some performance cost
Performance	KM may require more powerful hardware
Cost	Additional hardware to process signatures may incur a cost
Manageability	KM may require additional management overhead for managing certificates or key exchange
Usability	Usability of system resources may be affected by overhead of verifying signatures

Consequences. See Table 3.

Known Uses. A possible set of API for a version of the KM pattern is described AUTOSAR Key Manager [5].

Related Patterns. The KM patter is strictly related to both SCC pattern at the level of CAN (FD) communication and SSD pattern for the Automotive Ethernet communication. Also, it can cooperate with the HbIV pattern to guarantee the integrity and authenticity of the communications. Referring to [7], KM can be considered as a pre-condition of the *Symmetric Encryption* pattern. In addition, it manages keys for any kind of encryption both symmetric, asymmetric, and hybrid approaches.

2.4 Network-based Intrusion Detection (NbID)

Intent.

This pattern provides a mechanism to detect anomalous behavior in (automotive) network communication by observing metadata and extracted payload features from ECU messages.

Motivation. Communication within a modern vehicle is more and more connected to the outside world, which provides attackers with a multitude of new approaches to interfere and manipulate the internal communication within the vehicle network. To detect these potential manipulations, known intrusion signatures, as well as other unknown anomalies in data traffic, a dedicated system is required that can classify and verify the integrity of messages and is able to adapt to new situations.

Properties. The NbID can influence the integrity as well as the authentication property for observed network components, e.g. ECUs.

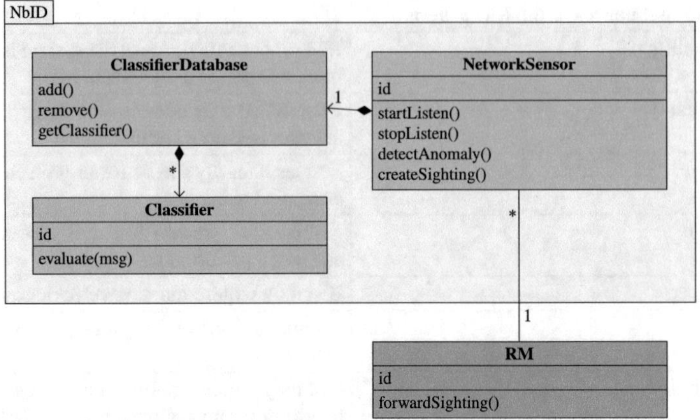

Fig. 9 Structure of the NbID pattern

Applicability. The NbID is applicable to attack detection. When integrated with an appropriate system, it can also be used for attack prevention and mitigation by triggering actions based on classification results. The pattern on its own does not apply to attack prevention or mitigation.

Structure.

A Network Sensor, as depicted in Fig. 9, is assigned to a specific network or component on a network to observe all send or received data from said component. The Classifier Database provides a set of Classifier models used to identify anomalies on the relevant system. The Classifier may be any type of either machine learning model [12, 31] or rule-based model [8] depending on the specific requirements of the current subsystem. Determining factors on what type of model to use may be the complexity of the expected data or the computation capabilities [1] of the component the NbID is deployed on. Classifiers are specifically created for the requirements of their relevant Network Sensor. The Sensor may select any combination of available Classifiers appropriate for its current situation. On the observation of an anomalous message, the network sensor is able to forward detection events. The RM accumulates detection events from the assigned Network Sensors and forwards sightings to a central reporting unit within the same network or an external SOC. One instance of the RM may be used for multiple Network Sensor instances on the same physical component.

Behavior. As general behavior of the NbID is shown in Fig. 10. The Network Sensor is observing the assigned network for new incoming or outgoing messages. As a new message is read, the respective classifiers are retrieved from the Classifier Database and used to analyze the message. Depending on the classifier result, a detection event is raised to the RM. Regardless of the classification results, the NbID does not manipulate or suppress any of its observed messages.

Constraints. In a real time environment, such as an automotive network, the classification of messages is required to be fast and efficient and must not influ-

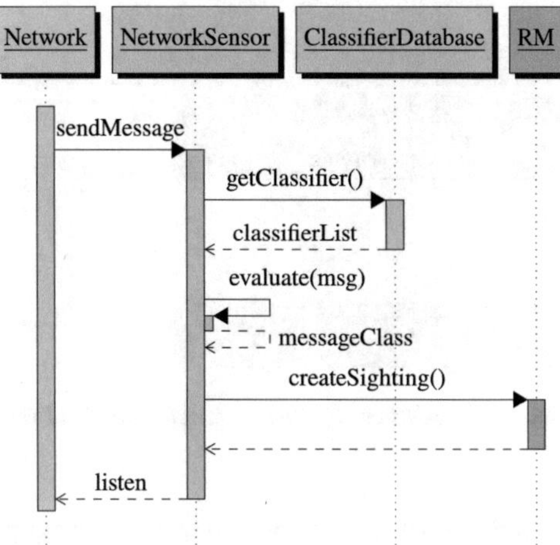

Fig. 10 Behavior of the NbID pattern

ence the performance or frequency of the message flow. Complex machine learning evaluations of every observed package are often not possible to perform within real time constrains. This is especially the case using neural network models without the use of dedicated hardware acceleration components, such as a Tensor Processing Unit (TPU), that may increase the cost of the component significantly. The available resources are therefore a limiting factor to the complexity of the classification process.

Consequences. Table 4 describes the consequences.

Table 4 Consequences for the NbID pattern

Accountability	Depends on type of IDS
Confidentiality	Not addressed
Integrity	Can be improved if ECU impersonation is detected
Availability	Might be improved if, e.g. as the result of classification, a prevention system can delete Denial of Service (DOS) messages can early, but is reduced by the overhead of classification
Performance	Classification cost and performance is depending on specific algorithms and hardware
Cost	Additional hardware, such as TPUs, to classify traffic may incur a cost
Manageability	RM controls sensors and accumulates all intrusion information
Usability	General safety requirements such as hard real time requirements and encryption might influence usability of the pattern

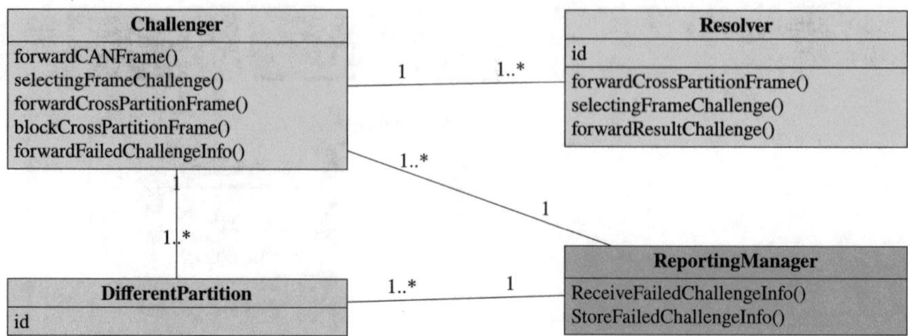

Fig. 11 Structure of *Challenge-based Intrusion Prevention (CbIP)* pattern

Known Uses. This pattern can be used similarly to the AUTOSAR specifications in [2, 3].

Related Patterns. The *Security Event Reporting* pattern is directly related and can be directly integrated. This pattern is also related to the Signature-based IDS pattern proposed by Cheng et al. [7] and can be integrated alongside to increase the detection rate of anomalies.

2.5 Challenge-based Intrusion Prevention (CbIP)

Intent. In an in-vehicle network, an unauthorized or compromised ECU could deliver messages in its partition without being discovered. CbIP pattern prevents this threat using knowledge challenges.

Motivation. Modern vehicles have to be considered fully connected to external IT systems like the road infrastructure and the web. This configuration exposes vehicles to different security and safety threats like the modification or the injection of malicious information among the ECUs. This pattern prevents an attacker could send malicious Controller Area Network (CAN) frames among different partitions of the in-vehicle network.

Properties. CbIP can assure two properties: the integrity of the CAN messages and the authentication of the sender.

Applicability. CbIP can be implemented in every in-vehicle CAN network and it can prevent and mitigate attacks like *fuzzing* and *replay*.

Structure. Figure 11 shows the structure of CbIP pattern. In particular, the two main actors of the in-vehicle network are a sender ECU, called the Resolver, and a partition Gateway (GW), called the Challenger, that enforces the prevention mechanism. To send a frame via Challenger to the destination, the Resolver must first successfully answer a specific question of the Challenger. To have a successful challenge, a sender ECU must correctly answer the challenge, and only under this condition the CAN frame is forwarded to the destination partition.

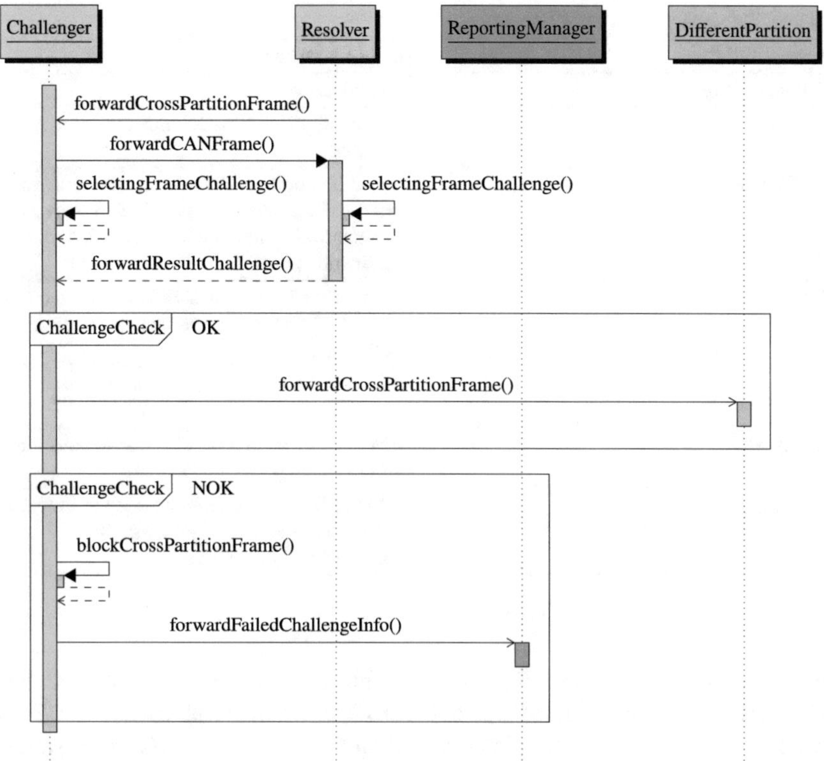

Fig. 12 Behavior *Challenge-based Intrusion Prevention (CbIP)* pattern

Behavior. CbIP monitors all the *cross-partition* CAN frames that are generated from an ECU belonging to a partition, e.g., untrusted partition, and addressed to one or more ECUs on another partition, e.g., trusted partition. The pattern can mitigate the attacks by verifying the authenticity of the Resolver by means of a challenge method. Thus, each Resolver has to know first, the *Database of CAN messages* (DBC), which is proprietary to a particular OEM and can be considered as a long-term secret only known by the CbIP and legitimate ECUs. Second, the *challenge set* represents the type of challenges that the CbIP will ask the ECU to authenticate. Third, the *encoding generation method* defines how frames, needed for the challenge, must be generated by the ECU that sent a cross-partition frame.

The Challenger demands the Resolver to solve a specific challenge, sending a CAN frame, as part of the handshake protocol (Fig. 12). If the Resolver correctly answers the challenge, for example, sending the correct frame of the DBC asked by the Challenger, this one has authenticated the Resolver ECU.

Only legitimate Resolvers know the full DBC, the set of challenges, and the encoding method. For this reason, only legitimate Resolvers can provide the correct answer. In this case, the message of the Resolver is authorized to be forwarded to the

Table 5 Consequences of the CbIP pattern

Accountability	Depends on type of CbIP
Confidentiality	Not addressed
Integrity	Can be improved if ECU impersonation is detected
Availability	Might be improved if one or more challenges fail in case of an attack like DoS. if detected, the involved frames are not forwarded to the destination partition
Performance	CbIP runs on low-power ECUs, however, more powerful ECUs can improve the challenge performance
Cost	Additional hardware with improved performances may incur a cost
Manageability	RM controls frames sent by ECUs and stores all failed challenge information
Usability	General safety requirements, such as strict real time requirements, can influence the application of the pattern

original cross-partition frame. If the answer is incorrect, the frame is discarded. If the Resolver has only partial knowledge of the DBC, it will not be able to rightly answer the challenge, since the pattern requires full knowledge of the DBC. Moreover, the Resolver that is not able to correctly answer the challenge may be banned from the list of authorized ECUs to send cross-partition frames. Thus, the Challenger may decide to discard any CAN frame sent by the banned ECU. Finally, CbIP foresees that the answer to the challenge is also sent to the RM to take a trace of the possible attacks.

Constraints. In a real in-vehicle network, challenging the ECU has to be made in a fast and efficient way to not affect safety.

Consequences. See Table 5.

Known Uses. Bella et al. [6] is an example of a possible implementation and application of CbIP.

Related Patterns. Both *Reporting Manager* and *NbID* patterns are related to CbIP. Besides, CbIP is related to the *Firewall* and the *Multi-Factor Authentication* pattern in [7].

2.6 Secure CAN Communication (SCC)

Intent. The target of this pattern is to assure secure communications among the in-vehicle ECUs to mitigate possible security threats.

Motivation. In-vehicle protocols were not designed to be fundamentally secure. For instance, the CAN bus protocol allows broadcasting messages in clear among all the ECUs in a specific partition of the in-vehicle network. The lack of security by design can expose communications to different threats like eavesdropping, man in the middle attack, or packet injection.

Properties. Over the in-vehicle communications, it is possible to apply different security mechanisms to assure authentication, data integrity, and confidentiality.

Applicability. The SCC pattern can assure the previous properties, generating frames to send among the different, ECUs that can prevent threats like spoofing or replay.

Structure. Figure 13 shows the structure of the SCC pattern. In an in-vehicle network, all ECUs can generate a secure frame to send to the other ECUs. Besides, ECUs can create *vanilla* protocol frames to guarantee backward compatibility.

In the SCC pattern, we have four main actors: the sender ECU, the receiver ECU, the Freshness Value Manager (FVM) that is in charge to generate and manage the freshness values, and the Crypto Service Manager (CSM) that manages the security keys. The three main elements to generate a secured frame are: (1) a MAC for authentication, (2) the FV for integrity, and (3) the encryption for confidentiality. Moreover, the SCC pattern algorithms were developed to not overload the timing and performances of the communication, for example assuring real-time communications do not affect safety.

The SCC security properties can be downgraded, for instance, it is possible to remove the confidentiality or the integrity mechanism to improve performance, but causing the loss of the related property like confidentiality or integrity. This possibility was maintained to apply lightweight versions of the pattern in case, for instance, fewer security guarantees are needed.

Behavior.

Figure 14 shows the behavior of the SCC pattern. Firstly, before sending the message, the ECU sender generates the MAC starting from the payload and the *Freshness Value* (FV) calculated according to the Monotonic Counter of the Freshness Manager as designed in Fig. 14. However, if necessary, the FV can be ignored by the ECU. The secured CAN frame is composed of the payload, the truncated MACT, and optionally, the truncated freshness value (FVT). Then it is encrypted and sent to the partition. When the ECU receives the CAN frame, before accepting and reading the message, it has to decrypt the CAN frame and verify the MAC to validate the message authenticity. The verification of the MAC occurs by using the received payload and a freshness value for verification (FVV) generated by the receiver, starting from the Monotonic Counter received by the FVM. Then, it computes the MAC using the received payload and the FVV. If the outcome equals the received MAC, then the payload is accepted, otherwise, it is rejected.

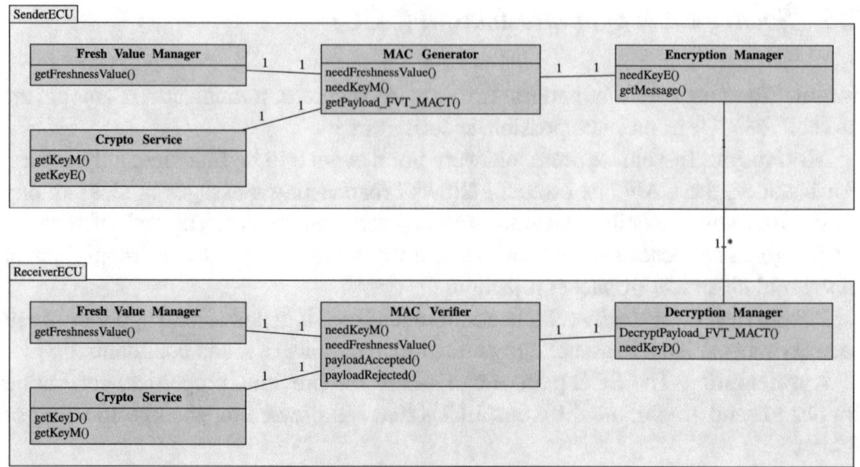

Fig. 13 Structure of the SCC pattern between two ECUs

Note that while MAC operations are normally fast, hence not problematic on inexpensive ECUs, it can be anticipated that the implementation of encryption and decryption primitives may cause a computational bottleneck. In consequence, the choice of the cryptographic scheme will have to be made with care. Usually, the Encrypt-then-MAC scheme is preferred, while this solution proposes the MAC-then-encrypt approach which has the following benefits in the CAN protocol: 1. The risk of message rebuilding is zeroed because there is no padding effect due to the fixed length of the considered messages and the used encryption algorithms with 64-bit block size; 2. The transmission of an additional frame to contain the MAC is not needed in case of frames where the 64 bits are already taken.

Constraints. To work, the SCC pattern needs to reduce the dimension of the payload of the message to introduce security information such as the MAC and the freshness value. However, this solution can lead to an increase of the traffic network, especially in the case of complex messages.

Consequences. See Table 6.

Known Uses. A possible design and deployment of a basic software module called CINNAMON, based on this pattern, is described in [6]. Besides, it has been designed according to the guidelines depicted by the AUTOSAR classic platform to assure integration with the in-vehicle systems. A lightweight version of the SCC pattern is the AUTOSAR SecOC software module [4]. However, SecOC does not assure the confidentiality of the frame to be lightweight.

Related Patterns. The pattern SCC is strongly related to the KM patter at the level of CAN (FD) communications. At this purpose, the KM can act as key generation and manager patter for the SCC pattern. Referring to [7], SCC can be considered an extension of the *Symmetric Encryption* pattern.

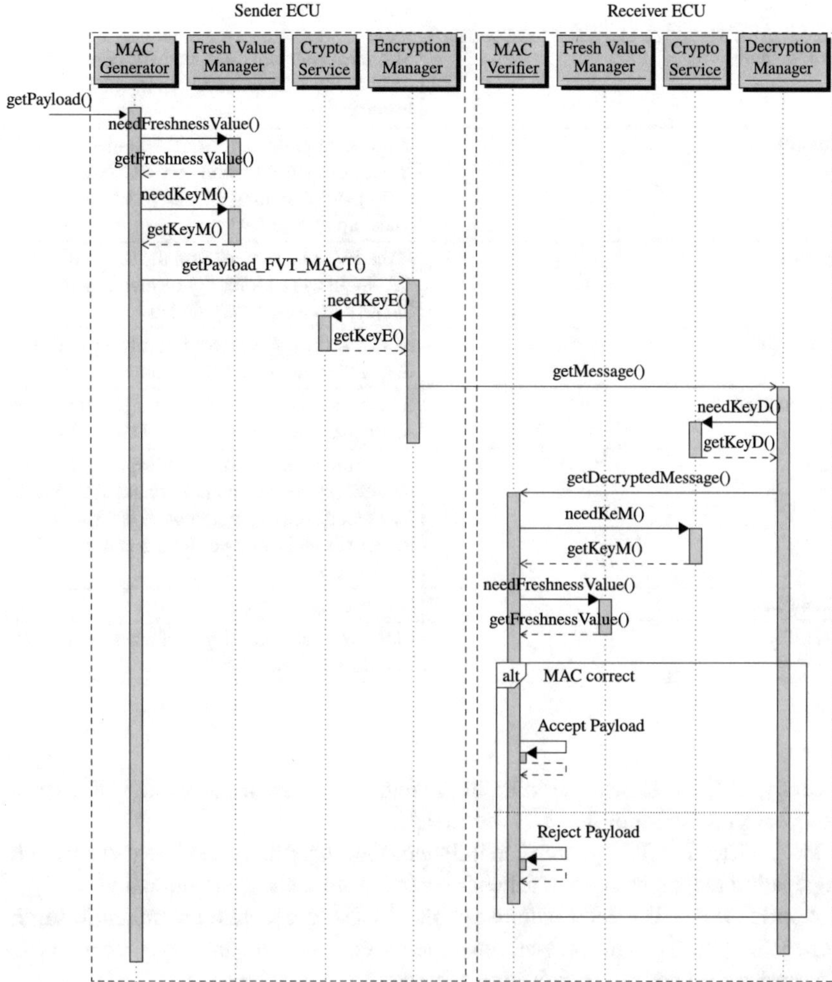

Fig. 14 Behavior of the SCC pattern between two ECUs

2.7 *Secure Service Discovery (SSD)*

Intent. Propagation of services of in-vehicle ECUs enable a service-oriented communication in a vehicle network. In-vehicle communication is key for the collaboration between. The goal of this pattern is to guarantee the authenticity of service propagation messages.

Motivation. Service-oriented communications become more and more relevant in vehicular networks, especially in automotive Ethernet. AUTOSAR introduced SOME/IP as well as DDS to enable service orientated communication between ECUs.

Table 6 Consequences of the SCC pattern

Accountability	SCC is accountable for integrity, confidentiality and senders authentication in its maximum instantiation
Confidentiality	If the encryption algorithm is applied, the CAN frames exchanged among the ECUs will be encrypted, providing data confidentiality on the transmitted payload
Integrity	If the FV is introduced into the secured frame, CAN will use an integrity mechanism to identify payload manipulation
Availability	SCC does not prevent availability, even if it may add some performance cost
Performance	It is influenced by the applied algorithm in the security pattern and by the ECU's hardware
Cost	The computational cost of the application of the security method can be considered low, however a significant hardware can improve the results and avoid possible communications bottlenecks
Manageability	Not directly addressed
Usability	Low overhead regarding performance may be introduced to the system

Recently, [37] has shown that man in the middle attacks are possible if the service discovery process is not properly secured.

Properties. The SSD pattern can influence the authenticity and integrity of service propagation messages and the following service-oriented communication.

Applicability. The SSD pattern enables ECUs to generate an authentic service propagation process. This prevents unauthentic entities to impersonate a regular ECU in a service-oriented communication schema.

Structure. The structure of the SSD pattern is depicted in Fig. 15. Every ECUs that offers services in the vehicle network must sign each service propagation message send to prevent attacks on service-oriented communication. In particular, the communication happens between service providers and service consumers. The service provider itself has a list of services it offers. Moreover, it has access to a Signature-Generater that has a connection with the FreshValueManager and the CryptoService. The CryptoService signs the message, and the FreshValueManager provides a freshness value.

The service consumer on the other hand also has a list of services it consumes. Furthermore, it has access to the SignatureVerifier that has a connection to the Fresh-ValueManager and the CryptoService. The CryptoService verifies the cryptographic signature and the FreshValueManager checks if the freshness of the provided value is acceptable.

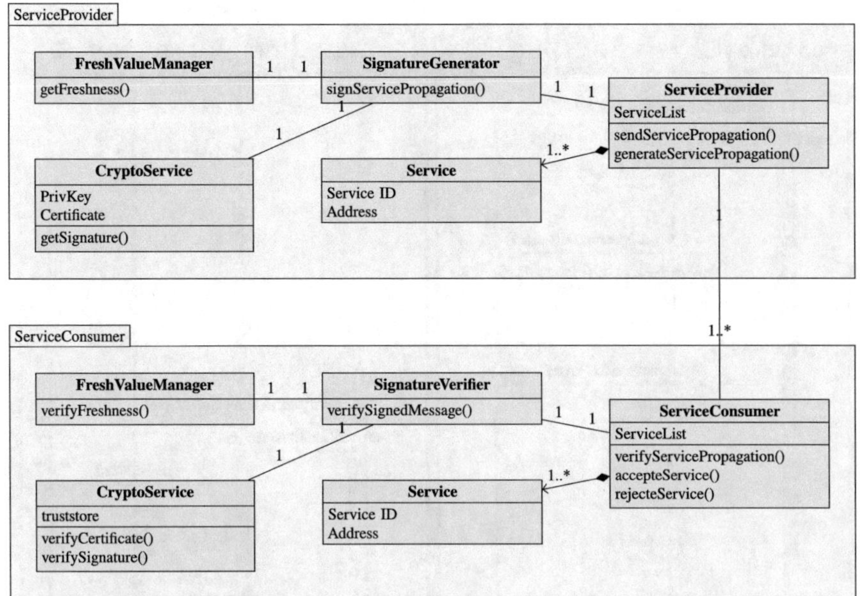

Fig. 15 Structure of the SSD pattern between two ECUs

Behavior. The behavior of the SSD pattern is depicted in Fig. 16. A service provider SP regularly generates new service propagation messages. These messages are signed by the Signature generator that extends the message by a freshness value from the fresh value manager. Then the crypto service signs the message and freshness value and attaches the service certificate.

The resulting message is sent to all service consumers. Every service consumer checks the correctness of the service propagation message with the signature verifier. The Verifier first let the Fresh value Manager check the freshness value. If this is correct, it lets the crypto service verify the signature of the message and controls the certificate. The certificate needs to contain the correct service ID as well as the being signed with the root of trust. If and only if all checks are correct, the service consumer updates its internal list of services and accepts the service provider.

Constraints. The current pattern design makes usage of asymmetric signatures for the authentication of service propagation messages. However, other approaches use symmetric cryptography, which requires less resources. These approaches also introduce the problem of a more complex key distribution process and keys for the authentication are known by multiple ECUs.

Consequences. Table 7 describes the consequences.

Known Uses. Different papers suggest the use of authentication schemes for service propagation of SOME/IP SD. Iorio et al. [17] and Zelle et al. [37], both suggest a scheme to sign a service discovery message and exchange a symmetric key for the further messages. Additionally, [37] suggest a second scheme using symmetric

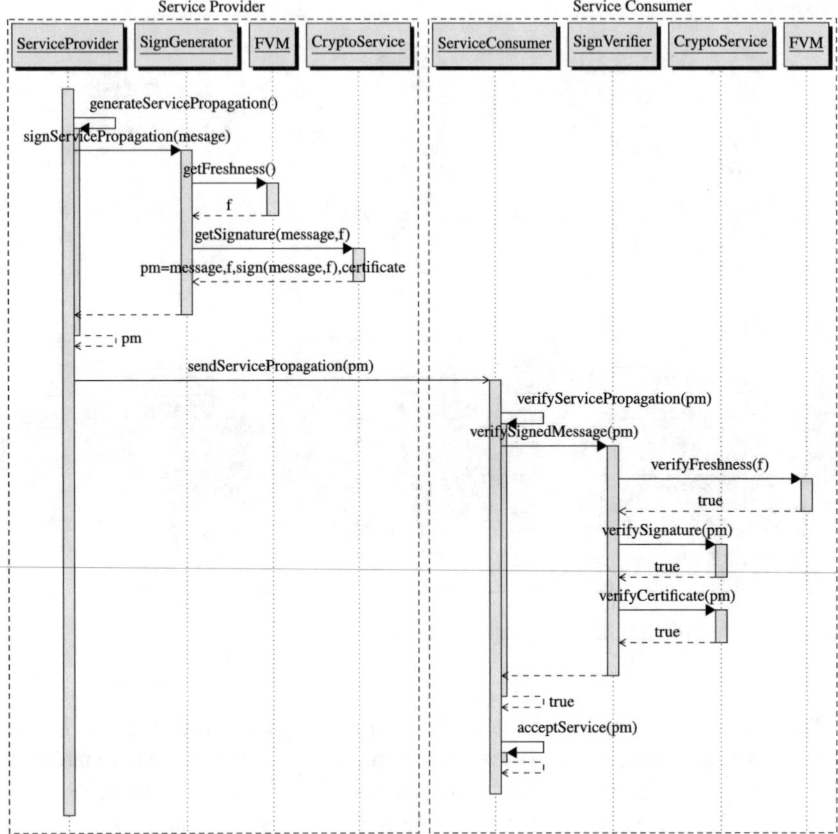

Fig. 16 Behavior of the SSD pattern between two ECUs

keys for the authentication against a central entity. Ma et al. [20] has a very similar scheme to this symmetric approach.

Additionally, the Object Management Group (OMG) Data Distribution Service (DDS) comes with a security extension to sign service propagation messages. This is for example described in [23].

2.8 Secure Feature Activation (SFA)

Intent.

This pattern allows authorized users to securely activate (previously purchased) features, e.g., general access to the vehicle, enhanced navigation maps, or online services, in the vehicle. The pattern guarantees authenticity and integrity properties.

Table 7 Consequences for the SSD pattern

Accountability	Not addressed
Confidentiality	Not addressed
Integrity	Integrity of the service propagation message is improved since it is cryptographically signed
Availability	Is not affected, since the cryptographic operations cannot be triggered externally
Performance	The ECU needs to perform additional cryptographic operations (sign and verify) thus the pattern introduces performance overhead
Cost	May need additional hardware cost for a more performant processor
Manageability	The PKI introduces a good manageability since it remove some key distribution issues
Usability	The signing of service propagation messages is not a standard feature for any implementation of DDS or SOME/IP

A backend issues fine-granular policies to authorized users that the user stores on a personal device. The policies contain the available features of the user. As soon as the user gets near the respective vehicle, the user can send the policies without an online connection via Bluetooth or NFC to the function controller of the vehicle. The function controller verifies the policies and activates the features.

The pattern is instantiated with a TPM as trust anchor in the vehicle that is attached to the TCU of the vehicle. Both TPM and TCU then constitute the function controller that is responsible to activate features across the vehicle. We completely map the pattern concept to the capabilities of the TPM. Thus, even a compromise of the TCU does not affect the security guarantees of the system.

The pattern allows implementing a majority of automotive use cases (UCs) that we clustered into 1. UC1: Local Feature Activation, 2. UC2: Remote Feature Activation, and 3. UC3: Online Feature Activation. With UC1, a feature of the TCU where the TPM is directly attached is activated. This could be for example be done by using the policy to unlock a decryption key that then decrypts encrypted map data, e.g., enhanced map data of Europe. With UC2, a feature within a remote ECU is activated. For example, a Keyed-Hash Message Authentication Code (HMAC) key could be unlocked by a policy, that could then be used in a challenge-response scheme to activate the feature in the remote ECU, e.g., an unlock of the doors within the door controller. UC3 is an online feature activation use case. Here, a policy unlocks an authentication key that can be used in a Transport Layer Security (TLS) handshake to establish a secure connection to the backend. An application for this would be the secure transmission of OTA updates, e.g., to implement a secure update mechanism according to UNECE 156, or the synchronization of music playlists.

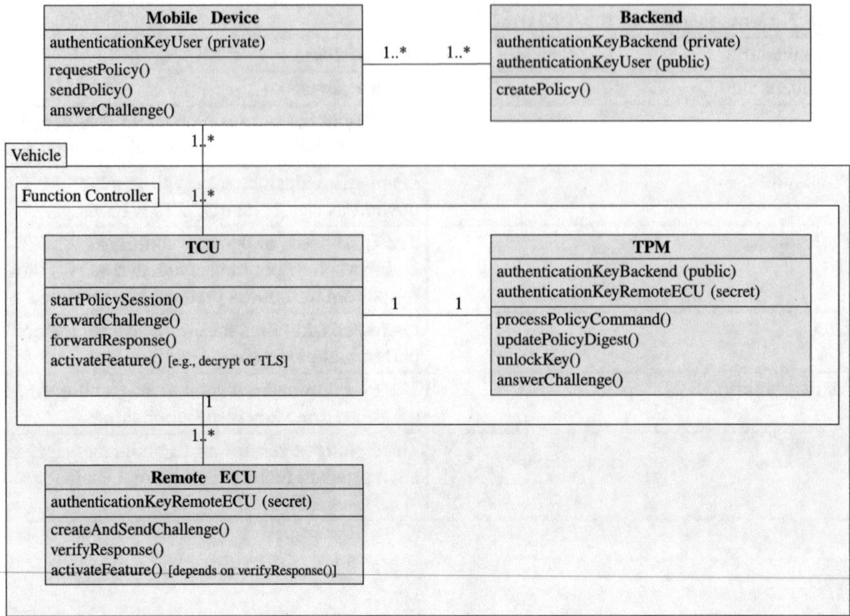

Fig. 17 Structure of the SFA pattern

Motivation. Feature activation is a modern business model of the OEMs to bind users to their brand. The business model will gain even more in importance with the oncoming of autonomous vehicles, where actual car ownership is decreasing. However, also the motivation of cyberattacks is high, since an illegal activating of the features guarantees high financial revenue.

Properties. The SFA pattern can be used to satisfy the integrity and authenticity properties of the feature activation business model.

Applicability. The SFA pattern enables to securely deploy the feature activation concept with respect to integrity and authenticity properties. Since it is fully mapped to the capabilities of the TPM, it also protects against physical and runtime attacks.

Structure.

The SFA pattern structure is depicted in Fig. 17. It roughly consists of the vehicle, the mobile device of the user, and the backend system. The important parts in the vehicle are the TCU with attached TPM and the remote ECU.

Behavior. The behavior of the SFA pattern in shown in Fig. 18 and highlights the actual authentication to activate the features. First, a user registers with the backend with the user public authentication key and optionally purchases certain features. At a later point in time the user requests the feature activation policy. Then the backend creates the according TPM policy, sends it to the mobile device, where it is stored. The TPM policy is at least cryptographically bound to the user and the backend. This can be achieved by configuring the TPM to only accept policies signed by the backend (backend binding) and to issue policies that require user interaction in a challenge-response scheme (user binding). Additionally, the TPM policy could also contain other restrictions like time or usage limitations.

Without an Internet connection needed, the user can activate the corresponding features by sending the TPM policies to the function controller of the vehicle. The TCU of the function controller will then start a policy session with the TPM and sends the policy commands one by one to the TPM. Within this policy processing loop, the user binding policy command will occur and a challenge response scheme with the user authentication key on the mobile device is triggered. If the verification was successful, the other policy commands are processed. If all policy commands could successfully be executed, the TPM checks the signature (backend binding) and grants key access.

The unlocked key is the feature activation key that can be used in the three previously mentioned Use Cases (UCs). In UC1, the feature key is a decryption key that may be used to decrypt some confidential data, e.g., enhanced map data. In UC2, the feature key is a HMAC key. A remote ECU starts a challenge-response scheme that is only successful if the challenge is authenticated with the correct HMAC key. In this case, the remote ECU activates the feature. In UC3, the feature key is an authentication key that is used in a TLS handshake to establish a secure connection to the backend, e.g., to initiate a secure OTA update, e.g., according to UNECE 156, or synchronize playlists.

Constraints. The current pattern design makes usage of asymmetric signatures for both backend and user binding, which is quite slow on the TPM [25]. However, the TPM also allows symmetric schemes, which would gain performance but would introduce the typical key distribution problem.

Consequences. Table 8 describes the consequences.

Known Uses.

This pattern makes extensive use of the TPM policies and is based on [25]. TPM policies are also used in other related work, e.g., in [14] where they are used to implement a secure OTA update mechanism or in [15] where they are used to select relevant features in unified firmware images. Work with a more generic feature activation scenario as well as the more specific feature activation use case "vehicle access" in car sharing scenario is detailed in [9, 10].

Related Patterns. The authorization, multifactor authentication, and tamper resistance patterns are related to the SFA pattern and partially address similar security goals [7].

3 Design Principles for Cyber Resilience

We now show how these patterns can be leveraged in design principles for cybersecurity engineering. The ISO/SAE 21434 standard for road vehicle cybersecurity engineering [18] recommends in its recommendation RC-10-06 that established and trusted design and implementation principles should be applied to avoid or minimize the introduction of weaknesses and it further refers to the design principles for archi-

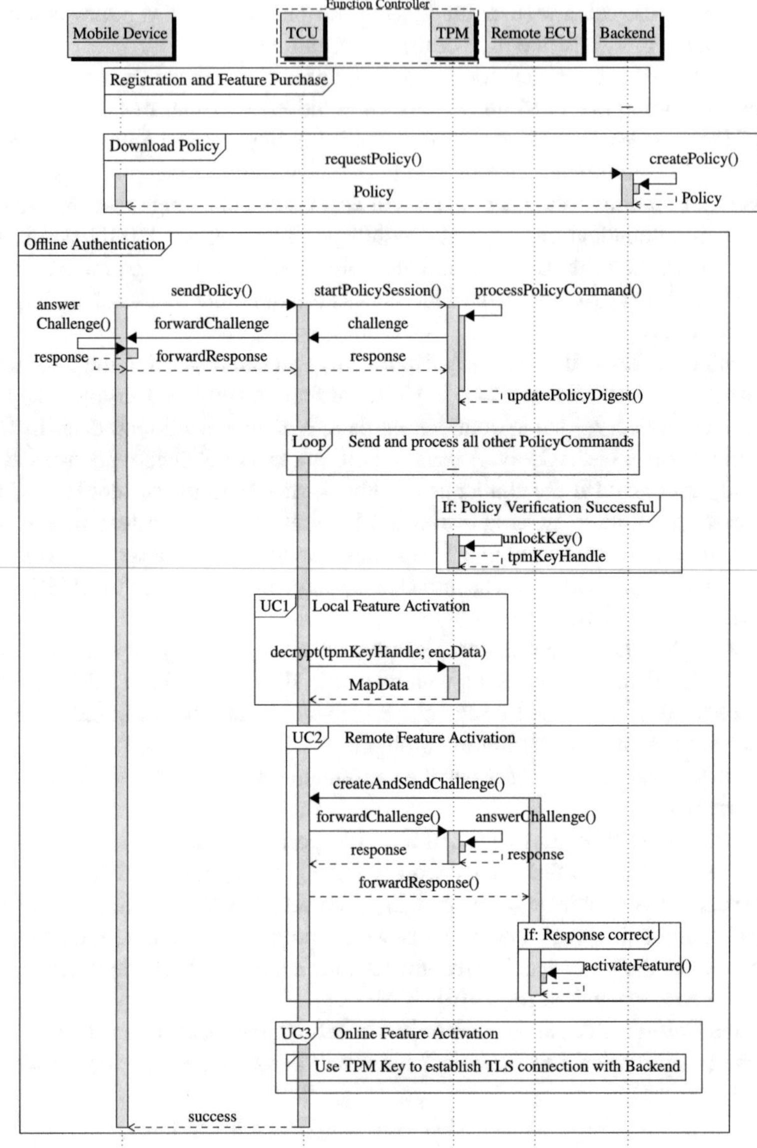

Fig. 18 Behavior of the SFA pattern

tectural design for cybersecurity in NIST Special Publication 800-160 Volume 1 [28]. While [28] addresses system security engineering in general, the recently published Volume 2 of NIST Special Publication 800-160 [29] focuses specifically on cyber resilience, with cyber resilience being described as "the ability to anticipate, withstand,

Table 8 Consequences for the SFA pattern

Accountability	Accountability is improved since the TPM establishes trust in the system
Confidentiality	Not directly addressed
Integrity	Integrity of the system is improved since the concept is completely mapped to the capabilities of the TPM
Availability	May be affected due to TPM performance overhead
Performance	The TPM has no cryptographic accelerators, thus the pattern introduces performance overhead
Cost	The TPM instantiation has additional hardware costs
Manageability	Standardized API and extensive software ecosystem facilitate management
Usability	The TPM chip introduce some overhead regarding performance and cost to the system

recover from, and adapt to adverse conditions, stresses, attacks, or compromises on systems that use or are enabled by cyber resources". In Table 9, we map our collaborative security patterns to these cybersecurity design principles to enable the development of trusted, secure automotive systems. Please note that some principles such as layered defense are achieved by combining the patterns.

4 Collaborative Patterns Applied in an Automotive Reference Architecture

In this section, we show an example of a concept for collaborative application of the proposed security patterns for robust attack detection and mitigation in a reference architecture. The reference architecture presented in Fig. 19 is based on modern domain-based E/E architectures structured into functional domains as proposed in [26]. We need a high level of detail to show where the patterns can be used within the architecture. The resource-constrained domain subnets are connected to a high-throughput backbone network, in this case based on Automotive Ethernet, via the gateway ECUs.

We placed the HbIV patterns at highly critical ECUs with external interfaces. For ECUs that offer the largest attack surface, e.g., connected to the Internet, the more secure TPM-based HbIV instantiation is chosen, while for the other ECUs the DICE-based HbIV instantiation seems sufficient.

Table 9 Mapping the patterns to the NIST structural cyber resiliency design principles [29]

Principle	Description	HbIV	KM	NbID	CbIP	SER	SCC	SSD	SFA
Limit the need for trust	Reduce the set of trusted entities in order to reduce the attack surface and lower assurance costs	✓	–	–	–	–	–	✓	✓
Control visibility and use	Visibility can be reduced by encryption/segmentation; use can be controlled by access control/least-privilege/need-to-know	✓	✓	–	–	–	✓	–	✓
Contain and exclude behaviors	Behaviors can be excluded, e.g., by time windows or dynamically. Containment can be achieved via predefined privileges and segmentation/isolation	✓	✓	✓	–	✓	–	✓	✓
Layer defenses and partition resources	Layering of defenses limits adversary vertical movement in a security architecture, while partitioning with controlled interfaces limits adversary lateral movement	✓	✓	✓	✓	✓	✓	✓	✓
Plan and manage diversity	Avoids the risk of system homogeneity. Because it can also increase the attack surface, cost, and risk of inconsistencies, the analysis of the use of diversity is crucial	✓	–	✓	–	–	–	–	✓
Maintain redundancy	Redundancy can improve the availability of critical functions and avoid single points of failure. To avoid spreading malware across homogeneous resources, it should be applied in conjunction with diversity	–	–	–	–	–	–	✓	–
Make resources location-versatile	Resources tied to specific locations can become single points of failure	–	–	–	–	✓	–	✓	–
Leverage health and status data	Health and status data supports situational awareness, alerts to potentially suspicious behavior, and can predict the need for adjustments	✓	–	✓	–	✓	–	✓	–
Maintain situational awareness	The detection of emerging anomalies supports decisions about measures to ensure safety and security goals	✓	–	✓	–	✓	–	–	–
Manage resources (risk-) adaptively	Support agility and help mitigate risk during critical operations despite component failures	–	–	–	✓	–	–	–	–
Maximize transience	Transient system elements minimize exposure time to adversary activity, while regularly updating them to a known (safe) state can wipe out malware or corrupted data	✓	–	–	–	–	–	–	–
Determine ongoing trustworthiness	Periodic or ongoing verification of the integrity of data and code, and ongoing analysis of the behavior of users, components, and services increases attacker effort and triggers responses	✓	–	✓	–	✓	–	✓	–
Change or disrupt the attack surface	Disruption of the attack surface can cause wrong adversary decisions	–	–	–	–	–	–	–	–
Make deception and unpredictability effects user-transparent	Unpredictable characteristics force the attacker to develop a broader range of techniques, but make it difficult to detect suspicious behavior	–	–	–	–	–	–	–	–

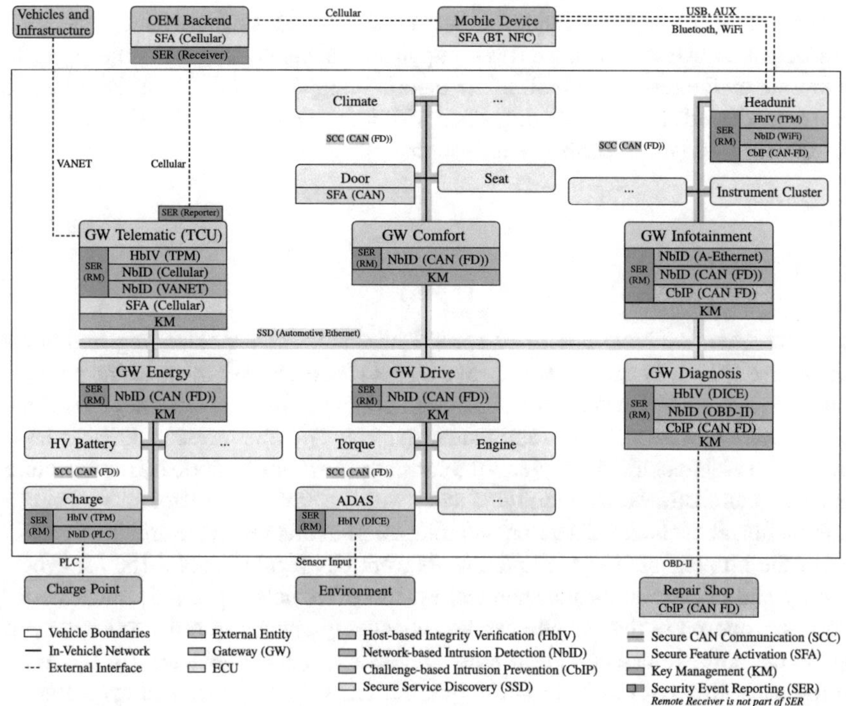

Fig. 19 Automotive reference architecture with collaborative observation and mitigation patterns

The KM pattern is placed into each gateway since it is in charge of providing the key to allow secured communication at the level of both Automotive Ethernet in collaboration with the SSD pattern and CAN (FD) communication in collaboration with the SCC pattern.

The NbID pattern is strategically placed to cover all communication channels in and to the outside of the vehicle.

The CbIP pattern is deployed on gateways that regulate possible cross-partition CAN messages and on the ECUs that need to know the challenge mechanism in order to respond correctly.

The SCC pattern is mainly placed on the bus since it aims to secure the communication on the bus by coding and decoding CAN messages according to the pattern description.

The SSD pattern is mainly placed on the Automotive Ethernet backbone since it aims to secure the service-oriented communication between gateways by authenticating service propagation messages.

According to the pattern structure (Fig. 17), the SFA pattern is deployed 1. in the backend, where the policies are issued, 2. on the mobile device of the user, where the policies are downloaded and stored, 3. on the TCU, where the TPM is attached, and 4. for example on the door controller as remote ECU to open the vehicle doors.

The patterns HbIV, NbID, and CbIP make use of the SER pattern to aggregate and distribute their events. Thus, the RM component that aggregates the events needs to be present on the respective ECUs. The reporter component that sends the events to the backend needs to be placed only once within the vehicle at the TCU since this is the primary interface to the backend systems.

5 Summary

In this chapter, we have described several new patterns for extending high-level automotive E/E architectures to improve attack resilience. We mapped the patterns to the NIST structural cyber resiliency design principles to show how they enable the development of trusted, secure automotive systems. We then presented a high-level automotive architecture that captures the currently discussed generic E/E architecture concepts at an abstract level and augmented it with the proposed collaborative security patterns for robust attack detection and mitigation in this reference architecture.

For the implementation of requirements from UN regulation No. 155 for cybersecurity and cybersecurity management system of vehicle types [35], the security patterns proposed in this chapter can be collectively applied in order to implement (a) the measures to detect and prevent cyberattacks against vehicles, (b) the monitoring capability with regard to detecting threats, vulnerabilities and cyberattacks relevant to the vehicle type, and (c) data forensics to enable analysis of attempted or successful cyberattacks.

In particular, the prevention of cyberattacks is supported by the KM pattern to exchange and store keys for encryption of intra-vehicle messages, the CbIP pattern prevents communication anomalies by challenge based authentication of ECUs, the SCC pattern secures the in-vehicle communication, the SSD pattern provides a secure mechanism to propagate and find service offers in an in-vehicle network, and, the SFA pattern provides a secure mechanism for authorized users to activate their (purchased) features in the vehicle.

Monitoring and threat detection capabilities are provided by the SER pattern which can aggregate security relevant events, such as software anomalies or unexpected network traffic, the HbIV pattern which can detect unauthorized modifications in ECU software and firmware at boot time, while the NbID pattern can detect anomalies in automotive network communication during driving.

Furthermore, data forensics to enable analysis of attempted or successful cyberattacks can be supported by the SER pattern.

Several requirements from UN regulation No. 156 on software update and software updates management system [36] could also be addressed by collaborative use of our security patterns described in this chapter. The HbIV pattern can be used to secure the update process by verifying the installed software state, the SFA pattern could be used to implement OTA updates, the SCC pattern would guarantee integrity/authenticity of updates during transport, while the KM pattern could manage the distribution of update keys.

6 Bibliographic Notes

The research in this field is diverse. As an example, the ENISA report on best practices for smart car security [11] lists a number of technical security measures that should be implemented to protect both smart cars and the associated back-end systems. General work on generic security design patterns has been undertaken for many application domains. However, research applying such a pattern to automotive systems is rare. In [21], Martin et al. propose a pattern-based approach that links protection and security patterns and provides guidance on selecting and combining both types of patterns. They demonstrate the application of these patterns by an automotive case study, however, very few examples are given. In [7], Cheng et al. describe a collection of security design patterns aimed at the automotive sector. They use an earlier security pattern template from [19] that is tailored for the development of secure systems, to which they extend automotive-specific fields. We also used this template in this chapter. They illustrate the applicability of their approach using a collection of ten automotive security design patterns.

Acknowledgements This work has been partly funded by the German Federal Ministry of Education and Research (BMBF) and the Hessen State Ministry for Higher Education, Research and the Arts within their joint support of the National Research Center for Applied Cybersecurity ATHENE and by the BMBF projects VITAF (ID 16KIS0835) and SAVE (ID 16KIS1324). Additionally, the project leading to this application has received funding from the European Union's Horizon 2020 research and innovation programme under grant agreement No 883135 (E-Corridor).

References

1. Al-Jarrah OY, Maple C, Dianati M, Oxtoby D, Mouzakitis A (2019) Intrusion detection systems for intra-vehicle networks: a review. IEEE Access 7:21266–21289. https://doi.org/10.1109/ACCESS.2019.2894183
2. AUTOSAR (2020) Specification of intrusion detection system manager for adaptive platform. https://www.autosar.org/fileadmin/user_upload/standards/adaptive/20-11/AUTOSAR_SWS_AdaptiveIntrusionDetectionSystemManager.pdf. Accessed 07 Oct 2021
3. AUTOSAR (2020) Specification of intrusion detection system protocol. https://www.autosar.org/fileadmin/user_upload/standards/foundation/20-11/AUTOSAR_PRS_IntrusionDetectionSystem.pdf. Accessed 07 Oct 2021
4. AUTOSAR (2020) Specification of secure onboard communication—CP Release 20-11. https://www.autosar.org/fileadmin/user_upload/standards/classic/20-11/AUTOSAR_SWS_SecureOnboardCommunication.pdf. Accessed 07 Oct 2021
5. AUTOSAR (2021) Specification of key manager. https://www.autosar.org/fileadmin/user_upload/standards/classic/21-11/AUTOSAR_SWS_KeyManager.pdf. Accessed 03 Oct 2022
6. Bella G, Biondi P, Costantino G, Matteucci I (2020) CINNAMON: a module for AUTOSAR secure onboard communication. In: 16th European dependable computing conference, EDCC 2020, Munich, Germany, 7–10 September 2020. IEEE, pp 103–110. https://doi.org/10.1109/EDCC51268.2020.00026

7. Cheng BHC, Doherty B, Polanco N, Pasco M (2020) Security patterns for connected and automated automotive systems. J Automot Softw Eng 1:51–77. https://doi.org/10.2991/jase.d.200826.001
8. Chevalier Y, Rieke R, Fenzl F, Chechulin A, Kotenko I (2019) ECU-secure: characteristic functions for in-vehicle intrusion detection. In: International symposium on intelligent and distributed computing. Springer, pp 495–504
9. Dmitrienko A, Plappert C (2017) Secure free-floating car sharing for offline cars. In: Proceedings of the seventh ACM on conference on data and application security and privacy, CODASPY '17. Association for Computing Machinery, New York, NY, pp 349–360. https://doi.org/10.1145/3029806.3029807
10. Dmitrienko A, Sadeghi AR, Tamrakar S, Wachsmann C (2012) Smarttokens: delegable access control with NFC-enabled smartphones. In: Katzenbeisser S, Weippl E, Camp LJ, Volkamer M, Reiter M, Zhang X (eds) Trust and trustworthy computing. Springer Berlin Heidelberg, Berlin, Heidelberg, pp 219–238
11. ENISA (2019) ENISA good practices for security of smart cars. https://www.enisa.europa.eu/publications/smart-cars. Accessed 07 Oct 2021
12. Fenzl F, Rieke R, Chevalier Y, Dominik A, Kotenko I (2020) Continuous fields: enhanced in-vehicle anomaly detection using machine learning models. Simul Model Pract Theory 105:102143. https://doi.org/10.1016/j.simpat.2020.102143
13. Fuchs A, Birkholz H, McDonald I, Bormann C (2021) Time-based uni-directional attestation. https://datatracker.ietf.org/doc/html/draft-birkholz-rats-tuda-04 (Work in Progress)
14. Fuchs A, Krauß C, Repp J (2016) Advanced remote firmware upgrades using TPM 2.0. In: Hoepman JH, Katzenbeisser S (eds) 31st IFIP International information security and privacy conference (SEC), vol AICT-471. Part 7: TPM and Internet of Things. Ghent, Belgium, pp 276–289. https://doi.org/10.1007/978-3-319-33630-5_19. https://hal.inria.fr/hal-01369561
15. Fuchs A, Krauß C, Repp J (2017) Runtime firmware product lines using TPM2.0. In: di Vimercati SDC, Martinelli F (eds) 32th IFIP International conference on ICT systems security and privacy protection (SEC). ICT systems security and privacy protection, vol AICT-502. Part 4: Operating system and firmware security. Springer International Publishing, Rome, pp 248–261. https://doi.org/10.1007/978-3-319-58469-0_17. https://hal.inria.fr/hal-01649005
16. Infineon (2019) A safe for sensitive data in the car: Volkswagen relies on TPM from Infineon. https://www.infineon.com/cms/en/about-infineon/press/market-news/2019/INFATV201901-030.html. Accessed 27 July 2021
17. Iorio M, Buttiglieri A, Reineri M, Risso F, Sisto R, Valenza F (2020) Protecting in-vehicle services: security-enabled SOME/IP middleware. IEEE Veh Technol Mag 15(3):77–85. https://doi.org/10.1109/MVT.2020.2980444
18. ISO/IEC (2021) ISO/SAE FDIS 21434—road vehicles—cybersecurity engineering
19. Konrad S, Cheng BHC, Campbell LA, Wassermann R (2003) Using security patterns to model and analyze security requirements. In: Heitmeyer C, Mead N (eds) 2nd International workshop on requirements engineering for high assurance systems (RHAS '03)
20. Ma B, Yang S, Zuo Z, Zou B, Cao Y, Yan X, Zhou S, Li J (2022) An authentication and secure communication scheme for in-vehicle networks based on SOME/IP. Sensors 22(2). https://doi.org/10.3390/s22020647. https://www.mdpi.com/1424-8220/22/2/647
21. Martin H, Ma Z, Schmittner C, Winkler B, Krammer M, Schneider D, Amorim T, Macher G, Kreiner C (2020) Combined automotive safety and security pattern engineering approach. Reliab Eng Syst Saf 198:106773. https://doi.org/10.1016/j.ress.2019.106773
22. Metzker E (2020) Reliably detecting and defending against attacks—requirements for automotive intrusion detection systems. VECTOR
23. Michaud MJ, Dean T, Leblanc SP (2018) Attacking OMG data distribution service (DDS) based real-time mission critical distributed systems. In: 2018 13th International conference on malicious and unwanted software (MALWARE), pp 68–77. https://doi.org/10.1109/MALWARE.2018.8659368
24. Plappert C, Fenzl F, Rieke R, Matteucci I, Costantino G, Vincenzi MD (2022) SECPAT: security patterns for resilient automotive E/E architectures. In: 2022 30th Euromicro international conference on parallel, distributed and network-based processing (PDP)

25. Plappert C, Jäger L, Fuchs A (2021) Secure role and rights management for automotive access and feature activation. Association for Computing Machinery, New York, NY, pp 227–241. https://doi.org/10.1145/3433210.3437521
26. Plappert C, Zelle D, Gadacz H, Rieke R, Scheuermann D, Krauß C (2021) Attack surface assessment for cybersecurity engineering in the automotive domain. In: 2021 29th Euromicro international conference on parallel, distributed and network-based processing (PDP), pp 266–275. https://doi.org/10.1109/PDP52278.2021.00050
27. RATS Working Group (2020) TPM-based network device remote integrity verification. https://datatracker.ietf.org/doc/html/draft-ietf-rats-tpm-based-network-device-attest-00. Accessed 13 July 2021
28. Ross R, McEvilley M, Oren JC (2018) Systems security engineering: considerations for a multidisciplinary approach in the engineering of trustworthy secure systems. Technical report, NIST Special Publication 800-160, vol 1, U.S. Department of Commerce, Washington, D.C. https://doi.org/10.6028/NIST.SP.800-160v1
29. Ross R, Pillitteri V, Graubart R, Bodeau D, McQuaid R (2021) Developing cyber-resilient systems: a systems security engineering approach. Technical report, NIST Special Publication 800-160, vol 2, Revision 1, U.S. Department of Commerce, Washington, D.C. https://doi.org/10.6028/NIST.SP.800-160v2r1
30. Swiderski F, Snyder W (2004) Threat modeling. Microsoft professional. Microsoft Press
31. Taylor A, Leblanc S, Japkowicz N (2016) Anomaly detection in automobile control network data with long short-term memory networks. In: 2016 IEEE International conference on data science and advanced analytics (DSAA), pp 130–139
32. Trusted Computing Group (2018) Hardware requirements for a device identifier composition engine. https://trustedcomputinggroup.org/resource/hardware-requirements-for-a-device-identifier-composition-engine/. Accessed 13 July 2021
33. Trusted Computing Group (2018) Implicit identity based device attestation. https://trustedcomputinggroup.org/resource/implicit-identity-based-device-attestation/. Accessed 13 July 2021
34. Trusted Computing Group (2019) TPM 2.0 library specification. https://trustedcomputinggroup.org/resource/tpm-library-specification/. Accessed 13 July 2021
35. UN Regulation No. 155 (2021) Uniform provisions concerning the approval of vehicles with regards to cyber security and cyber security management system. https://unece.org/sites/default/files/2021-03/R155e.pdf [Online]. Accessed 30 Apr 2021
36. UN Regulation No. 156: Uniform provisions concerning the approval of vehicles with regards to software update and software updates management system (2021). https://unece.org/sites/default/files/2021-03/R156e.pdf [Online]. Accessed 30 Apr 2021
37. Zelle D, Lauser T, Kern D, Krauß C (2021) Analyzing and securing SOME/IP automotive services with formal and practical methods. In: The 16th international conference on availability, reliability and security, ARES 2021. Association for Computing Machinery, New York, NY. https://doi.org/10.1145/3465481.3465748

Explainability of Model Checking for Mobile Malicious Behavior Between Collaborative Apps Detection and Localisation

Francesco Mercaldo, Rosangela Casolare, and Antonella Santone

Abstract The technological development of recent years has made possible to improve the performance of mobile devices such as smartphones, tablets, smart TVs and wearable devices. This improvement has introduced the possibility of developing more complex applications able to manage sensitive user data. An example is represented by banking applications: they allow us to carry out all the financial operations that we can perform in a physical bank. Therefore, it is clear that in order to carry out these operations, we need a very high level of security to be sure that attackers do not use our account to transfer our money. Users typically install applications on their smartphones, without checking the required permissions before installation, because they do not know what risks they can encounter. Among the various malicious attacks that can be perpetrated, the collusive attack is emerging, as threat targeting devices based on the Android operating system. In this attack paradigm, two or more apps collaborate in some way to perform a malicious action that they are unable to do independently. Detection of colluding apps is a challenging task, as a matter of fact free and commercial antimalware analyse each app separately, hence fail to detect any joint malicious action performed by multiple collaborative apps through collusion. The contribution of this paper is a proposal of an explainable technique exploiting model checking, aimed to localise the malicious instructions in the application under analysis, by automatically identifying the bytecode instructions performing a malicious collusion and, for this reason, making the proposed method explainable.

F. Mercaldo (✉) · R. Casolare · A. Santone
University of Molise, Campobasso, Italy
e-mail: francesco.mercaldo@unimol.it

R. Casolare
e-mail: rosangela.casolare@unimol.it

A. Santone
e-mail: antonella.santone@unimol.it

T. Dimitrakos et al. (eds.), *Collaborative Approaches for Cyber Security in Cyber-Physical Systems*, Advanced Sciences and Technologies for Security Applications,
https://doi.org/10.1007/978-3-031-16088-2_5

1 Introduction

The evolution of communication devices due to the growing technological development of recent years has brought about a notable change in our daily habits.

Their diffusion was too fast due to the possibility of buying medium-high quality smartphone (or tablet) models on the market at affordable prices, another reason is characterized by the possibility of obtaining a large number of applications freely.

As a consequence of this rapid spread, it was not possible to provide adequate training to users, so that they could understand what was the correct use of the devices. Normally, users use smartphones without thinking about the issues associated with their use: these devices handle data of different types on a daily basis, including personal information which is the main target of cybercriminals [15].

There exist several operating systems in the mobile environment,in this context the most popular represented by Android and , for this reason, it represented the preferred target from malware writers [14, 20].

Applications that can be installed on devices can be downloaded from different parts: often people use official markets (i.e., Google Play) to download applications, but it is possible to consider unofficial markets (for instance, AppChina), so as to be able to download applications that require payment on the official store for free, another reason is given by the possibility of installing applications not available on the official Android market [11, 25].

Third-party markets are often unreliable, since the applications they offer can be infected, but it is not certain that all the applications on the official markets are reliable, in fact even in the latter there are malicious applications even if in smaller quantities [7, 13]. In this scenario we are considering there are two main actors: on the one hand we have the *attacker*, represented by cyber criminals whose purpose is to develop malicious code to attack users in order to exfiltrate their sensitive and private information, and on the other hand there are *defenders* represented by tools i.e., the antimalware, that have the ability to detect the presence of threats [16, 24]. However, antimalware are not able detect new types of threats, as they need threat signature to be able to recognize them, which must already be known and stored in their repository [3, 11].

Working with the aim of increasing the complexity of malicious payloads (i.e., the code that takes care of performing the malicious action) and being able to evade antimalware controls, malware authors have developed a new type of threat: the *Colluding Attack*. A collusive attack involves splitting the malicious action between two or more applications. By doing this, the antimalware that analyzes the single application is not able to detect the present threat, because is the communication between these applications the attack.

A typical colluding attack occurs in the following way: there are two infected applications, capable of communicating with each other, when the user takes a certain action or when a particular event occurs in the system. The first application has the task of reading the user sensitive data and then sending them to the second, which has the task of transmitting them to the outside world. In order to perform these actions, the first application has the permission to read the data, while the second has the permission to connect to a network, then sharing the information.

Inter-Component Communication (ICC) is a type of communication that allows Android applications to communicate with each other: this mechanism helps the developer in the implementation of communication mechanism, allowing them to take advantage of the reuse functionality [5]. The existence of this mechanism, however, means that the applications are not always independent of each other, allowing the collaboration between applications, requires that there is an exchange of information between components belonging to the same application or to different applications [29]. The latter option can be used by malicious applications to attack user data [8, 23].

Recently model checking [10, 12] was exploited with the aim of detecting and verifying the presence of collusive applications in Android environment [11]. In particular the method proposed in [10] uses a heuristic function aimed at reducing the number of applications candidates for analysis. To define the function, authors consider the μ-calculus temporal logic and model checking.

This work represents an extension of the paper entitled "Detecting Colluding Inter-App Communication in Mobile Environment" [10] published in the Applied Science journal.

The main features of the discussed approach for Android colluding detection are the following:

- the possibility of identifying colluding attacks relating to the three Android shared resources mentioned above;
- the possibility of automatically detecting a collusion, in fact there is an algorithm that allows you to automatically generate the properties that are used to detect the presence of a collision between applications;
- collision is detected even if it occurs between more than two applications.

We focus the attention on the possibility to perpetrate colluding attack exploiting *SharedPreferences*, *ExternalStorage* and *BroadcastReceiver*.

Each of these resources can be exploited to launch an attack, as explained below:

- using *SharedPreferences* we can store key-value pairs of data, configuration and preferences. It is therefore possible for an application to save some settings (containing information on user data) in a shared preferences file, which could then be read by the receiving application;
- using *ExternalStorage* it is possible to create a file in which to store information. For example, in this scenario, an application could read the data and send it to a second application by leveraging an intent. The second application then, using the external memory could send them in turn to the outside or to a third application;
- using *BroadcastReceiver* it is notified when an event occurs within the system (i.e., a receipt of SMS). It is useful for managing special events instantly. Broadcast receivers to respond to sent broadcast messages use Intent objects, either from the same application or from other applications installed on the device. Thanks to this

component it is possible to have applications that receive data through broadcast receivers and through SMS messages [5].

In particular in this chapter we add following contributions:

- we design an algorithm to explain the rationale behind the colluding detection performed by the model checker;
- we propose a method for the detection of the bytecode instructions responsible for the malicious payload: this can help the malware analysts to group different malware in the same malicious family and can be also considered as a step forward the sanitization of malicious behaviours;
- we show, by exploiting a real-world application, how the proposed algorithm can be considered for effective malicious payload bytecode instructions localisation.

In the next section we introduce preliminary notions about model checking, Sect. 3 present the proposed method for the explainable detection of colluding malicious behaviour in Android environment, the results of the experimental analysis are shown in Sect. 4, current state-of-the-art literature is discussed in Sect. 5 and, finally, in the last section conclusion and future research direction are drawn.

2 Background About Model Checking

This paragraph describes preliminary concepts on the model checking technique considered by the proposed approach. The first step to use the model checking technique is to represent a system using a formal specification language. In the example we are considering, the Calculus of Communicating Systems (CCS) is used in a slightly extended form. Systems are described in terms of processes. Processes perform actions and evolve, thus becoming new (and different) processes after each action.

Below we report the syntax used to describe the processes:

$$p ::= DONE \mid x \mid \alpha.p \mid p + p \mid p; p \mid p \| p \mid p \backslash L \mid p[f]$$

the variable α ranges over a finite set of actions $A = \{\tau, a, \overline{a}, b, \overline{b}, ...\}$. With the symbol τ we represent the internal action.

The set of visible actions V (ranged over by l), is defined as $A - \{\tau\}$. L is a set in processes of the form $p \backslash L$, that represents a set of actions such that $L \subseteq V$, instead f is the relabelling function in processes of the form $p[f]$, and it is a total function, $f : A \to A$, such that the constraint $f(\tau) = \tau$ is respected.

For each action $l \in V$ (resp. $\overline{l} \in V$) we have a complementary action $\overline{l}$ (resp. l); it holds that $\overline{\overline{l}} = l$.

Given $L \subseteq V$, with L^+ we denote the set $\{\overline{l}, l \mid l \in L\}$. The variable x ranges over a set of constant names: each constant x is defined by a constant definition $x \stackrel{\text{def}}{=} p$, where p is the *body* of x.

Table 1 Operational semantics for the extended calculus of communicating systems

Rule		Rule	
Done	$\dfrac{}{DONE \xrightarrow{\delta} nil}$	**Act**	$\dfrac{}{\alpha.p \xrightarrow{\alpha} p}$
Seq$_1$	$\dfrac{p \xrightarrow{\alpha} p'}{p; q \xrightarrow{\alpha} p'; q} \quad \alpha \neq \delta$	**Seq**$_2$	$\dfrac{p \xrightarrow{\delta} nil}{p; q \xrightarrow{\delta} q}$
Sum	$\dfrac{p \xrightarrow{\alpha} p'}{p + q \xrightarrow{\alpha} p'}$ (and symmetric)	**Par**	$\dfrac{p \xrightarrow{\alpha} p'}{p \Vert q \xrightarrow{\alpha} p' \Vert q} \quad \alpha \neq \delta$ (and symmetric)
Com$_2$	$\dfrac{p \xrightarrow{\delta} p',\ q \xrightarrow{\delta} q'}{p \Vert q \xrightarrow{\delta} DONE}$	**Com**$_1$	$\dfrac{p \xrightarrow{l} p',\ q \xrightarrow{\bar{l}} q'}{p \vert q \xrightarrow{\tau} p' \vert q'}$
Res	$\dfrac{p \xrightarrow{\alpha} p'}{p \backslash L \xrightarrow{\alpha} p' \backslash L} \quad \alpha \notin L^+$	**Rel**	$\dfrac{p \xrightarrow{\alpha} p'}{p[f] \xrightarrow{f(\alpha)} p'[f]}$
Con	$\dfrac{p \xrightarrow{\alpha} p'}{x \xrightarrow{\alpha} p'} \quad x \stackrel{\text{def}}{=} p$		

$DONE$ is the constant defined as $\delta.nil$ and represents a process whose task is to terminate the execution of δ action, in fact the process nil cannot perform any action and it is also named *deadlocked*. P denotes the set of all processes.

Given a set D of constant definitions, the standard operational semantics S is given by a relation $\longrightarrow_D \subseteq P \times A \times P$. $\longrightarrow_D$ is the least relation defined by the rules in Table 1. Given a process p the semantic of p is the automaton called *standard transition system* for p and is denoted $S(p)$.

If we wanted to informally explain the semantics of an extended CCS process, it can be seen that there is no rule for the nil process, so it cannot perform any action. In the *Done* rule, the process can execute δ and thus reach a deadlock state.

As we can see in *Act* rule, the $\alpha.p$ process can perform the α action becoming the process p.

Seq$_1$ and *Seq*$_2$ rules represent the sequencing of two processes. The q process can starts its execution only when the p process ends its execution performing the δ action.

Sum rule claims that p and q represent alternative choices for the behavior of $p + q$. The operator $\Vert$ expresses the parallel execution. *Par* rule instead, shows how processes in a parallel composition can behave autonomously: if the process p performs α and becomes p', then $p \Vert q$ performs α and becomes $p' \Vert q$.

When *Com*$_1$ rule is used, we say that a handshake occurs and in this case two processes simultaneously execute the complementary actions; the handshake results in an internal communication as the action τ. At the end of both p and q processes, the $p \| q$ process becomes $DONE$. The operator $\backslash L$, in *Res* rule, prevents actions in L^+ to be done: if p can perform α and becomes p', then $p \backslash L$ can perform α to become $p' \backslash L$ but only if $\alpha \notin L^+$.

In *Rel* rule, the operator $[f]$ can rename actions by means of the relabelling function f: so whether p can perform α to become p', then $p[f]$ can perform $f(\alpha)$ becoming $p'[f]$. Thus a constant x acts as p if $x \stackrel{\text{def}}{=} p$ as stated in rule *Con*, which states that a process behaves like its definition.

Once obtained the formal model of a system S it is necessary to prove properties of this S system. This is accomplished by using a temporal logic. In the case under analisys it is used *mu-calculus* logic, which syntax is below reported. We suppose that Z ranges over a set of variables, K and R range over sets of actions A.

$$\begin{aligned}\phi \ ::= & \ \texttt{tt} \mid \texttt{ff} \mid Z \mid \phi \vee \phi \mid \phi \wedge \phi \mid \\ & [K]\, \phi \mid \langle K \rangle\, \phi \mid \nu Z.\phi \mid \mu Z.\phi\end{aligned}$$

We can define as follows the satisfaction of a ϕ formula by state s of a transition system, denoted by $s \models \phi$:

- every state satisfies `tt` and no state satisfies `ff`;
- a state satisfies $\phi_1 \vee \phi_2$ ($\phi_1 \wedge \phi_2$) if it satisfies one (or both) ϕ_1 ϕ_2;
- $[K]\, \phi$ and $\langle K \rangle\, \phi$ represent the modal operators: $[K]\, \phi$ is satisfied by a state which, for every execution of an action in K, it changes in a state obeying ϕ; instead $\langle K \rangle\, \phi$ is satisfied by a state which can change to a state obeying ϕ by performing an action in K.

Table 2 shows the precise definition of the satisfaction of a closed formula φ by a state s (denoted $s \models \varphi$).

Table 2 Satisfaction of a closed formula by a state

$p \not\models \texttt{ff}$	and	$p \models \texttt{tt}$
$p \models \varphi \wedge \psi$	iff	$p \models \varphi$ and $p \models \psi$
$p \models \varphi \vee \psi$	iff	$p \models \varphi$ or $p \models \psi$
$p \models [K]\, \varphi$	iff	$\forall p'.\forall \alpha \in K.p \xrightarrow{\alpha} p'$ implies $p' \models \varphi$
$p \models \langle K \rangle\, \varphi$	iff	$\exists p'.\exists \alpha \in K.p \xrightarrow{\alpha} p'$ and $p' \models \varphi$
$p \models \nu Z.\varphi$	iff	$p \models \nu Z^n.\varphi$ for all n
$p \models \mu Z.\varphi$	iff	$p \models \mu Z^n.\varphi$ for some n

where:

– for each n, $\nu Z^n.\varphi$ and $\mu Z^n.\varphi$ are defined as:

$\nu Z^0.\varphi = \texttt{tt}$ $\qquad \mu Z^0.\varphi = \texttt{ff}$

$\nu Z^{n+1}.\varphi = \varphi[\nu Z^n.\varphi/Z] \quad \mu Z^{n+1}.\varphi = \varphi[\mu Z^n.\varphi/Z]$

where the notation $\varphi[\psi/Z]$ denotes the substitution of ψ for every free occurrence of the variable Z in φ.

$\mu Z.\phi$ and $\nu Z.\phi$ are the fixed point formulae, where μZ (νZ) *binds* free occurrences of Z in ϕ. An occurrence of Z is free if it is not within the scope of a binder μZ (νZ). While, a formula is *closed* if it contains no free variables.

$\mu Z.\phi$ is the minimum fixed point of the recursive equation $Z = \phi$, while $\nu Z.\phi$ is the maximum. T a transition system, satisfies a ϕ formula, defined $T \models \phi$, if and only if $q \models \phi$, where q is the state initial of T. A CCS p process satisfies ϕ s and $S(p) \models \phi$.

In the following definition we consider these abbreviations: K ranges over sets of actions and A is the set of all actions.

$$\begin{aligned} \langle \alpha_1, \ldots, \alpha_n \rangle \varphi &\stackrel{\text{def}}{=} \langle \{\alpha_1, \ldots, \alpha_n\} \rangle \varphi \\ \langle - \rangle \varphi &\stackrel{\text{def}}{=} \langle \mathcal{A} \rangle \varphi \\ \langle -K \rangle \varphi &\stackrel{\text{def}}{=} \langle \mathcal{A} - K \rangle \varphi \end{aligned}$$

Once defined the model and the temporal logic properties, it is necessary to have something enabling us to check if the model satisfies the defined properties. For this purpose, formal verification is considered, it is a system process that, exploiting mathematical reasoning, is able to verify whether a system (i.e., the model) satisfies certain requirements (i.e., the temporal logic properties).

Among the various verification techniques is the Model Checking technique. In it the properties are formulated in temporal logic and each property is evaluated against the system (i.e., the LTS-based model). The model checker accepts as input a model and a property, returning "TRUE" if the system satisfies the formula and "FALSE" otherwise. The verification performed is an exhaustive search in the state space, since the model is finite it is guaranteed to end.

In the proposed method we consider the Concurrency WorkBench of the New Century (CWB-NC)[1], a widespread formal verification environment. Note that we can easily use CWB-NC for extended CCS as it offers the ability to translate new operators using standard CCS ones.

3 Detecting Colluding Inter-App Communication

In this section we present the proposed method aimed to detect the collusive application of Android. This method is the first designed for the detecting different types of collusive attacks: *ExternalStorage*, *SharedPreferences* and *BroadcastReceiver*.

Figure 1 shows an overview about the proposed method.

Starting from an Android device, the applications installed on it are collected for analysis by obtaining the APKs of the applications at the url */data/app* in the internal memory of the device. It is possible to read these url without root permissions: the permissions on that directory are *rwxrwx–x*. The APK extension refers to an Android

[1] https://www3.cs.stonybrook.edu/~cwb/.

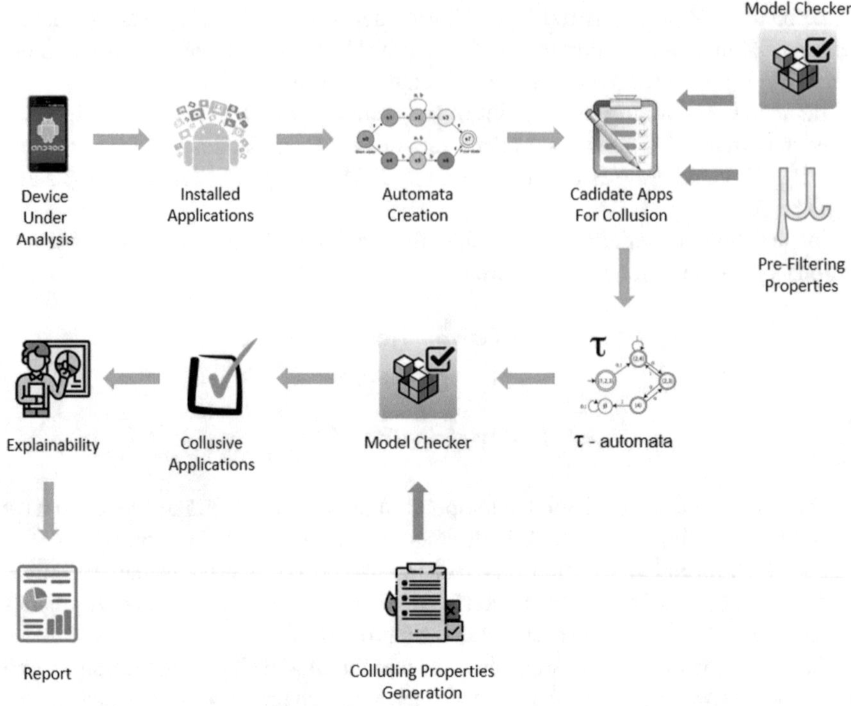

Fig. 1 The colluding properties generation step

package file. This type of file is a variant of the JAR file and is used for the distribution and installation of bundled components on the Android mobile platform. The APK has all the resources necessary for the running application (i.e., external libraries, images, audio and the set of application *.class* files) stored in the *classes.dex* file, in a format translated for the Dalvik Virtual Machine of Android.

Once the APKs of the *Installed Applications* have been collected, through a reverse engineering process it is possible to obtain the Java bytecode for each APK.
For each method of Android applications it is possible to have the Java bytecode representation using the *dex2jar*[2] tool. From the classes.dex file (the executable file for the Dalvik Virtual Machine of Android), it is possible to obtain the .jar file (the executable for the Java Virtual Machine) and the Byte Code Engineering Library (BCEL),[3] useful for having the Java bytecode representations of the classes stored in the .jar file.
Java bytecode is considered because it also allows obfuscated APKs to be obtained, for example using the R8 compiler integrated in the latest release of Android studio,[4]

[2] https://github.com/pxb1988/dex2jar.

[3] https://commons.apache.org/proper/commons-bcel/.

[4] https://developer.android.com/studio/build/shrink-code.

which provides some sort of obfuscation to shorten the names of classes and methods; but it also applies optimization techniques that can make the code more difficult to read, taking advantage of strategies to reduce the size of the developed application.

A model is generated for each Java bytecode relating to the installed APKs (i.e., *Automata Creation* step in Fig. 1). For the generation of the automata we consider a CCS model for each method of an APK and it is possible to obtain it by performing the translation of each Java bytecode instruction in a CCS process.
The translation of the sequential Java bytecode instructions is as follows:

$$proc\ x_{current} = opcode.x_{next}$$

here $x_{current}$ denotes the current instruction being parsed, instead x_{next} denotes the process representing the next instruction and *opcode* is the name of the Java bytecode instruction. The syntax of the New Century Concurrency Workbench, the formal verification environment used for experimentation, is used to express constant definitions. For this reason, instead of $x \stackrel{\text{def}}{=} p$ we write $proc\ x\ =\ p$.

In the Listings 5.1 and 5.2 an example of CCS translation from the translation of sequential instructions of the operative code is shown.

Listing 5.1 CCS process for Listing 5.1

```
proc P1 = store.M2
proc P2 = load.M3
proc P3 = return.nil
```

Listing 5.2 CCS process for Listing 5.2

```
proc P1 = dup.M2
proc P2 = invokesubstring.M3
proc P3 = pushConstant.M4
proc P4 = newStringBuilder.M5
proc P5 = invokeinit.M6
proc P6 = pop.M7
proc P7 = return.nil
```

In Listing 5.1 there are only three actions: store, load and return; in Listing 5.2 we have more actions, for example related to method invocations (i.e., requiresubstring and invokeinit), to the definition of new Java objects (i.e., newStringBuilder), to stack instructions such as pop (which allows to remove the value in top of stack) or dup (which allows to duplicate the value at the top of the stack), but also the push instruction (which allows to insert a variable onto the stack and in this case the variable is "Constant").

Branch instructions are used to change the instruction execution sequence. We consider the $+$ operator for choice management.

A CCS process is built for each method of the application under analysis. To better understand: let *aua* be an application under analysis. Assuming *aua* has n methods, which are $F_1, \ldots, F_n$; the *aua* CCS representation has n $M_1, \ldots, M_n$ CCS as processes.

The automata generated in terms of the CCS process are considered as inputs for the Model Checker, with a set of formulas for verifying whether the applications exhibit a GET or PUT operation relating to an Android resource of type *ExternalStorage*, *SharedPreferences* or *BroadcastReceiver*. A series of *Pre-Filtering Properties* are used for this verification. The automata that return TRUE from the Model Checker constitute *Candidate Apps For Collusion*.

Table 3 The table shows the properties: φ_{PUT} which allows you to detect methods that invoke PUT operations on $SharedPreferences$, and the property φ_{GET} which allows you to detect methods that invoke $GEToperations$ on $SharedPreferences$

$Formula\_SP\_PUT$	
φ_{PUT}	$= \mu X.\langle invokegetSharedPreferences\rangle\ \varphi_{PUT_1} \vee$ $\langle -invokegetSharedPreferences\rangle\ X$
φ_{PUT_1}	$= \mu X.\langle invokeedit\rangle\ \varphi_{PUT_2} \vee \langle -invokeedit\rangle\ X$
φ_{PUT_2}	$= \mu X.\langle invokeputString, invokeputInt, invokeputFloat\rangle\ \varphi_{PUT_3} \vee$ $\langle -invokeputString, invokeputInt, invokeputFloat\rangle\ X$
φ_{PUT_3}	$= \mu X.\langle invokecommit\rangle\ \mathtt{tt} \vee \langle -invokecommit\rangle\ X$
$Formula\_SP\_GET$	
φ_{GET}	$= \mu X.\langle invokegetSharedPreferences\rangle\ \varphi_{GET_1} \vee$ $\langle -invokegetSharedPreferences\rangle\ X$
φ_{GET_1}	$= \mu X.\langle invokegetString, invokegetInt, invokegetFloat\rangle\ \mathtt{tt} \vee$ $\langle -invokegetString, invokegetInt, invokegetFloat\rangle\ X$

As for *SharedPreferences*, *ExternalStorage* and *BroadcastReceiver*, an application can perform two types of operations on a shared resource: PUT and GET. For *SharedPreferences* the following actions are coded using the μ-calculus logic:

- for an application that performs a PUT action on a shared resource, the formula contained in Table 3—$Formula\_SP\ _PUT$ will be true if the following actions are performed: *invokegetSharedPreferences*, *invokeedit*, *invokeputString/invokeputInt/invokeputFloat*, *invokecommit*;
- for an application that performs a GET action on a shared resource, the formula contained in Table 3—$Formula\_SP\_GET$ will be true if the following actions are performed: *invokegetSharedPreferences*, *invokegetString/invokegetInt/invokegetFloat*.

The Table 4 shows χ_{PUT} and χ_{GET}, they are used to detect PUT and GET operations on *ExternalStorage* respectively.
For *ExternalStorage* the following actions are coded using the μ-calculus logic:

- for an application that performs a PUT action on a shared resource, the formula contained in Table 4—$Formula\_ES\_PUT$ will be true if the following actions are performed: *invokegetExternalStorageDirectory*, *invokewrite*;
- for an application that performs a GET action on a shared resource, the formula contained in Table 4—$Formula\_ES\_GET$ will be true if the following actions are performed: *invokegetExternalStorageDirectory*, *invokereadFully*.

The Table 5 shows ψ_{PUT} and ψ_{GET}, they allow you to detect PUT and GET operations on the *BroadcastReceiver* respectively. For *BroadcastReceiver* the following actions are coded using the μ-calculus logic:

- for an application that performs a PUT action on a shared resource, the formula contained in Table 5—$Formula\_BR\_PUT$ will be true if the following actions are performed: *invokeputExtra*, *invokesendBroadcast*;

Table 4 The properties: χ_{PUT} which allows to detect methods which invoke PUT operations on $ExternalStorage$, and the property χ_{GET} which allows to detect methods which invoke GET operations on $ExternalStorage$

$Formula\_ES\_PUT$	
χ_{PUT}	$= \mu X. \langle invokegetExternalStorageDirectory \rangle\, \chi_{PUT_1} \vee$ $\langle -invokegetExternalStorageDirectory \rangle\, X$
χ_{PUT_1}	$= \mu X. \langle invokewrite \rangle\, \mathtt{tt} \vee$ $\langle -invokewrite \rangle\, X$
$Formula\_ES\_GET$	
χ_{GET}	$= \mu X. \langle invokegetExternalStorageDirectory \rangle\, \chi_{GET_1} \vee$ $\langle -invokegetExternalStorageDirectory \rangle\, X$
χ_{GET_1}	$= \mu X. \langle invokereadFully \rangle\, \mathtt{tt} \vee$ $\langle -invokereadFully \rangle\, X$

Table 5 The table shows the properties: ψ_{PUT} which allows to detect methods which invoke PUT operations on $BroadcastReceiver$, and the property ψ_{GET} which allows to detect methods which invoke GET operations on $BroadcastReceiver$

$Formula\_BR\_PUT$	
ψ_{PUT}	$= \mu X. \langle invokeputExtra \rangle\, \psi_{PUT_1} \vee$ $\langle -invokeputExtra \rangle\, X$
ψ_{PUT_1}	$= \mu X. \langle invokesendBroadcast \rangle\, \mathtt{tt} \vee$ $\langle -invokesendBroadcast \rangle\, X$
$Formula\_BR\_GET$	
ψ_{GET}	$= \mu X. \langle invokegetStringExtra \rangle\, \psi_{GET_1} \vee$ $\langle -invokegetStringExtra \rangle\, X$
ψ_{GET_1}	$= \mu X. \langle invokestart \rangle\, \mathtt{tt} \vee$ $\langle -invokestart \rangle\, X$

- for an application that performs a GET action on a shared resource, the formula contained in Table 5—$Formula\_BR\_GET$ will be true if the following actions are performed: *invokegetStringExtra*, *invokestart*.

These properties were built with the idea of obtaining three different sets of applications, each related to a different collusion attack (i.e., *SharedPreferences*, *ExternalStorage* and *BroadcastReceiver*), for each set we have the classes (i.e., the CCS automata) tha are resulting correctly verified by the properties of *PUT* and *GET* for each collusion attack.

Model checking, considering as input a class modeled in terms of CCS and a logical temporal formula, performs a screening and selects only the potentially threatening classes for launching a colluding attack. This significantly reduces the number of classes to be tested and therefore also reduces processing costs.

After obtaining from the *Installed Applications* the set of applications that can potentially show collusive behaviors, we move on to the next steps illustrated in Fig. 2, which have the purpose of detecting whether two or more applications can effectively sharing information by making a colluding attack.

For each automaton belonging to the *Candidate Apps For Collusion*, a simplified version is generated containing only the action that can be involved in the collusive attack. In this regard, actions that cannot be involved in collusive attacks are replaced with τ actions.

Specifically, the models resulting from the τ-automata step contain the push action: it implies the presence of the read and write operation on a variable, as well as a set of actions that discriminate if the colluding attack occurs via *Shared-Preferences*, *ExternalStorage* or *BroadcastRecevier*. Starting from the τ-automata models, a set of properties is automatically generated to check for collusion between applications.

To better understand its construction, Listing 5.3 shows an example of an automaton relating to a PUT operation of a collusion attack using a *ExternalStorage* resource, and in Listing 5.4 it is shown an example related to the τ-automaton.

Listing 5.3 CCS process for Listing 5.3

```
proc M1 = invokeinit.M2
proc M2 = invokegetExternalStorage.M3
proc M3 = pushConstant1.M4
proc M4 = store.M5
proc M5 = pushConstant2.M6
proc M6 = load.M7
proc M7 = pushConstant3.M8
proc M8 = invokewrite.M9
proc M9 = return.nil
```

Listing 5.4 CCS process for Listing 5.4

```
proc M1 = t.M2
proc M2 = invokegetExternalStorage.M3
proc M3 = pushConstant1.M4
proc M4 = t.M5
proc M5 = pushConstant2.M6
proc M6 = t.M7
proc M7 = pushConstant3.M8
proc M8 = t.M9
proc M9 = t.nil
```

The automaton shown in Listing 5.3 gives *true* as a result to the property relating to the PUT of *ExternalStorage* (we refer to χ_{GET} in the Table 4). There is an $invokegetExternalStorage$ action and, after a certain number of actions, the $invokewrite$ action.

Listing 5.4 shows the automaton τ relative to the automaton shown in Listing 5.3. Here all actions are translated into τ actions, since they are not useful for collusion detection, except for actions involving a *push* type operation (in this example we find $pushConstant1$, $pushConstant2$ and $pushConstant3$, respectively aimed at pushing the variables Constant1, Constant2 and Constant3 into the stack) and $invokegetExternalStorage$ one.

It is worth remembering that for the GET and PUT automata, in addition to the push actions, the τ-automata consider the $invokegetExternalStorage$ action for the collusive attack *ExternalStorage*; for *SharedPreferences* we consider the $invokegetSharedPreferences$ action; for the *BroadcastReceiver* we consider $invokeputExtra$ and $invokeSendBroadcast$ actions (for the PUT), while $invokeStart$ and $invokeStringExtra$ (for the GET) and the push in analysis (*BroadcastReceiver* for both the properties of GET than for those of PUT).

After obtaining all the τ-automata for the GET and PUT operations of all three considered collusion attacks, the authors in [10] have designed an algorithm whose flow chart is represented in Fig. 2, for the automatic *Colluding Properties Generation*, specifically for the generation of *ExternalStorage*, *SharedPreferences* and *BroadcastReceiver* properties for collusion detection.

The algorithm was designed with the aim of automatically deducing a set of properties and, in case, a PUT and a GET automata (belonging to *SharedPreferences*, *ExternalStorage* or *BroadcastReceiver*) result true to the properties including the same push action, then the automata PUT and GET are marked as executors of a collusive attack.

Considering the flow diagram shown in Fig. 2, let us see in detail the steps of the algorithm to which it refers.
For each attack considered, that is *SharedPreferences*, *ExternalStorage* and *BroadcastReceiver* we have as input the set of τ-automata relating to the PUT operation. For each PUT automaton all the PUSH operations are taken, through a $sort$ operation, so as to show the set of visible actions of the automata[5]; for each action represented by a push operation, the rules involved in the push and the following actions are automatically generated:

- about *SharedPreferences* the $invokegetSharedPreferences$ and the pushes in analysis are considered;
- about *ExternalStorage* the $invokegetExternalStorage$ and the pushes in analysis are considered;
- about *BroadcastReceiver* the $invokeputExtra$, $invokeSendBoradcast$ (for the PUT property) and the $invokeStringExtra$, $invokeStart$ (for the GET property) are considered.

The properties are automatically generated based on the PUT τ-automata, as described in the flowchart in Fig. 1.

For example, referring to the PUT *ExternalStorage* τ-automaton in Listing 5.4the properties automatically generated by the proposed algorithm will be those shown in Table 6, where for each push action of the τ-automaton generates a property.

The process repeats for each PUT τ-automata push action considered. Once all the properties for all the push actions of all the τ-automata have been generated, they are stored in a file in the step *write generate properties* of the flowchart in Fig. 1.

After that, the model checker is invoked with the PUT and GET τ-automata and the properties obtained in the *Colluding Properties Generation* phase in Fig. 1; the

[5] http://courses.cs.vt.edu/cs5204/fall00/CWB/top-level-coms.html.

Table 6 Automatically generated properties for the τ automaton of Listing 5.4: ς_{push1} property is related to the $pushConstant1$ action, ς_{push2} property is related to the action $pushCostant2$ and ς_{push3} property is related to the $pushCostant3$ action

$Formula\_Constant1\_push$	
ς_{push1}	$= \mu X. \langle invokegetExternalStorageDirectory \rangle \, \varsigma_{push12} \vee$ $\langle -invokegetExternalStorageDirectory \rangle \, X$
ς_{push12}	$= \mu X. \langle pushConstant1 \rangle \, \mathtt{tt} \vee$ $\langle -pushConstant1 \rangle \, X$
$Formula\_Constant2\_push$	
ς_{push2}	$= \mu X. \langle invokegetExternalStorageDirectory \rangle \, \varsigma_{push22} \vee$ $\langle -invokegetExternalStorageDirectory \rangle \, X$
ς_{push22}	$= \mu X. \langle pushConstant2 \rangle \, \mathtt{tt} \vee$ $\langle -pushConstant2 \rangle \, X$
$Formula\_Constant3\_push$	
ς_{push3}	$= \mu X. \langle invokegetExternalStorageDirectory \rangle \, \varsigma_{push32} \vee$ $\langle -invokegetExternalStorageDirectory \rangle \, X$
ς_{push32}	$= \mu X. \langle pushConstant3 \rangle \, \mathtt{tt} \vee$ $\langle -pushConstant3 \rangle \, X$

properties are checked first on the PUT τ-automata, if the result is TRUE, then the verification is also performed with the GET τ-automata. If the same property is also TRUE on the GET τ-automata, there is collusion.
Collusive attack is possible between PUT and GET τ-automata and therefore between two (or more) applications, whose PUT and GET models return TRUE to the automatically generated property.

Once the collusion has been detected, the method returns as output a report indicating the applications involved in the collusion, their classes and methods that are related to the collusion, the name of the shared variable (i.e., push action) and the type collusion attack (i.e., *SharedPreferences*, *ExternalStorage*, *BroadcastReceiver*).

In Fig. 2 we report the workflow for the algorithm related to the explainability task.

The explainability task is aimed to provide more details with respect to the classification task provided by the model checking based classification. In detail, the aim of this task, exploiting the model checker, is to output the class, the method and the exact bytecode instructions responsible for the malicious colluding action. In this way the proposed method is able not only the output a label, indicating whether an application is able perform a colluding action, but also to automatically explain the malicious instruction, i.e., the reason why the application under analysis is able to perform the specific malicious action.

As shown from the workflow in Fig. 2, the designed algorithm takes two different input: the first one is a model resulting *TRUE* from the previous analysis and the relative temporal logic formula succesfully verified on this model. So, in the next steps we split the model in a series of submodels (as shown from the *Splitting model in*

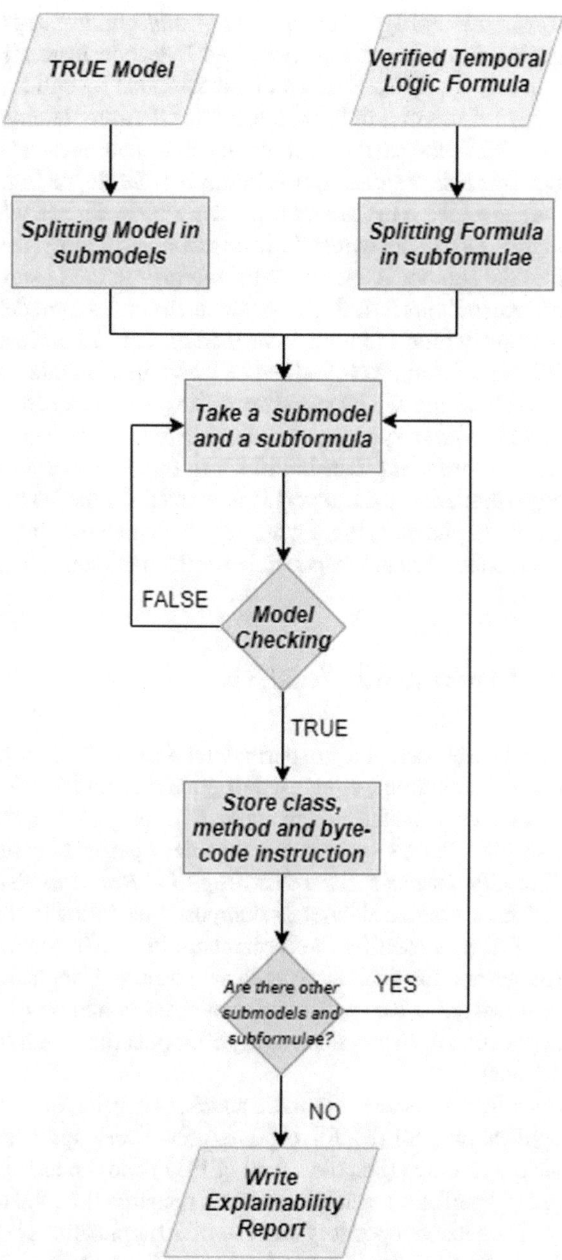

Fig. 2 The workflow of the proposed algorithm for the explainability task

submodels and *Splitting Formula in subformulae* steps in Fig. 2), where each model contains just one action (i.e., one bytecode instruction). By exploiting a similar process, we obtain a series of subformulae by splitting the verified temporal logic formula in a series of formulae, each of them containing just one bytecode instruction to verify. In the next we consider the first submodel and the first subformula (as shown from the *Take a submodel and a subformula* step in Fig. 2) to input the model checker (as shown from the *Model Checking* step in Fig. 2), whether the model checker will output *TRUE* the bytecode instruction verified by the subformula is implemented into the submodel and the bytecode instruction is stored otherwise (i.e., the model checker outputs *FALSE*), we evaluate the next submodels until the formula is verified. And every time a formula is verified on a model, its instruction is stored in bytecode. When a subformula is verified on a certain submodel, it obviously passes to the next formula which will be tested on the next submodel compared to the one that has been verified with the previous subformula (as shown from the *Are there other submodels and subformulae?* decision block in Fig. 2). Once all subformulae are verified, the algorithm ends and a report is generated (in the *Write Explainability Report* in the workflow shown in Fig. 2): this report contain the class, the method and the bytecode instructions that are responsible for the malicious collusion.

4 Experimental Analysis

The dataset used in the experimental analysis was built from several Android application repositories, composed of malicious and legitimate Android applications. In order to evaluate the effectiveness of the proposed method, it is necessary to consider a set of malicious applications capable of performing the collusive attacks of the type *SharedPreferences*, *ExternalStorage* and *BroadcastReceiver*.

The evaluated dataset is composed as follows: the first repository considered is *ACE*, proposed by the researchers in [5]. In this paper, the authors designed a framework for generating collusive Android applications. It is composed of 482 different collusive applications considering attacks of type *SharedPreferences* (160 applications), *ExternalStorage* (160 applications) and *BroadcastReceiver* (162 applications).

Specifically, for each collusive attack, consisting of 160 (162 for *BroadcastReceiver*) applications, 80 (81 for *BroadcastReceiver*) applications perform a write on the shared resource (i.e., the action of PUT) and the remaining 80 (81 for *BroadcastReceiver*) perform a read on the same resource (i.e., the action of GET).

The second repository consists of 20 applications developed by the authors in [9]. Since the ACE dataset contains *SharedPreferences* collusive applications, that share only string variables, in this dataset there are 10 applications that exploit a collusion attack through a *Int* value and 10 other applications with the collusion attack via a *Float* value.

Table 7 Malware families in the *Drebin* dataset. In the first column *Family* there are the 10 most populous families of this repository, in the column *Inst.* we have the type of delivery payload (standalone, repackaging, update), in the column *Attack* the type attack (trojan, botnet) and in the last column *Activation* there are the system events that activate the malicious payload

Family	Inst.	Attack	Activation
FakeInstaller	s	t, b	
DroidKungFu	r	t	boot, batt, sys
Plankton	s, u	t, b	
Opfake	r	t	
GinMaster	r	t	boot
BaseBridge	r, u	t	boot, sms, net, batt
Kmin	s	t	boot
Geinimi	r	t	boot, sms
Adrd	r	t	net, call
DroidDream	r	b	main

To verify the correctness of the approach when finding non-collusive applications (which must be classified as trusted), the repository *DroidBench* 2.0[6] is considered; this dataset consists of 120 non-collusive applications, but some of them use *Shared-Preferences*.

In addition, to evaluate the effectiveness of the approach in detecting collusive attacks only, the *Drebin* malware repository, composed of malicious Android applications [2, 6], which do not carry out collusive attacks, was chosen. 10 malware belonging to the top 10 malicious families in the dataset were selected, having a total of 100 non-colluding malware, as shown in Table 7.

Has been considered a set of 50 ransomware samples, obtained from AMD[7] repository. Specifically, the samples considered belong to the malicious *Simplelocker* family, whose payload encrypts files with different extensions on the device's external memory and then requests a ransom in bitcoin to decrypt the data.

Let's also consider a set of 300 trusted applications taken from the official Google store (i.e., Play Store). The applications considered were automatically collected by Google Play,[8] thanks to the use of a script developed by authors that is responsible for querying and downloading applications from the Android official market.
These 260 applications were scanned with the *VirusTotal*[9] service to confirm their trustworthiness. VirusTotal runs 60 different antivirus (i.e., Kaspersky, Avast, McAfee) on each application. At the end of the scan it was confirmed that the trusted applications present in the legitimate dataset do not contain malicious payloads.

---

[6] https://github.com/secure-software-engineering/DroidBench.

[7] http://amd.arguslab.org/.

[8] https://play.google.com/store.

[9] https://www.virustotal.com/.

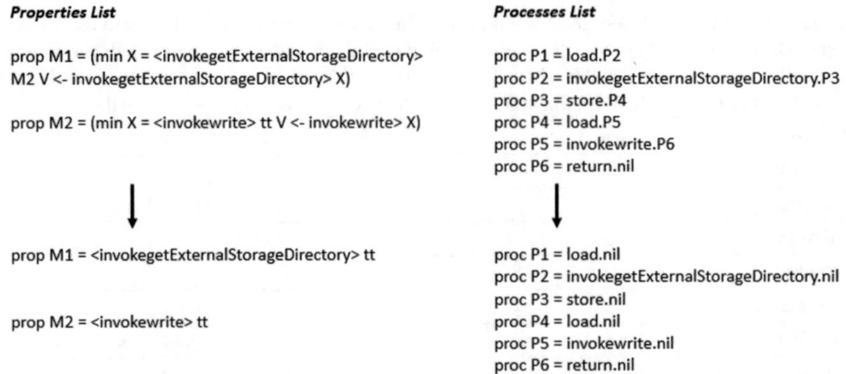

Fig. 3 Example of the structure of the list of processes and properties to which the explainability is applied

Therefore, a total of 1072 applications were collected: 772 malicious applications (collusive and otherwise) and 300 legitimate applications.

In Fig. 3 we show an example of how the splitting of the Model and Formula reported in the initial part of the workflow 2.

We have a column for the models (and submodels) of the Formula Properties and a column for the models (and submodels) of the Model Processes list.
Each line of the model is transformed in the submodel, into a statement that does not relate to another subsequent instruction, but ends after its execution. In this way, it was possible to create an "isolation" of the individual instructions for each process. Then for each isolated instruction (i.e., submodel, subformula) we check whether it is TRUE for the Model Checker: if the verification is successful (TRUE), then we go to save the method and move on to the next one; if the check returns a negative result (FALSE), then it goes to the next method until finds the method with which the collusion occurs (i.e., the Model Checker returns TRUE). If there is no TRUE between the methods, it means that they are not involved in a collusion.

4.1 Experimental Results

To measure and evaluate the performance of the approach, four different metrics were considered: Precision, Recall. F-Measure and Accuracy.

Precision allows to calculate the proportion of examples that are really part of class 'x', compared to all those assigned to it. To calculate it, we refer to the ratio between the number of relevant records retrieved and the total number of irrelevant and relevant records retrieved. The formula is as follow:

Table 8 Performance results

Precision	Recall	F-Measure	Accuracy
1	1	1	1

$$Precision = \frac{tp}{tp+fp}$$

The values *tp* and *fp* in the formula indicate the number of true positives and the number of false positive respectively.

Recall allows to calculate the proportion of examples assigned to class 'x', among all the examples that actually belong to the class; it says how much of the class has been captured. To calculate it, we consider the relationship between the number of relevant records retrieved and the total number of relevant records. The formula is as follow:

$$Recall = \frac{tp}{tp+fn}$$

The *fn* value in the formula indicates the number of false negatives.

F-Measure is used to measure the accuracy of a test. This value can be understood as a weighted average of Precision and Recall. The formula is as follow:

$$F\text{-}Measure = 2 * \frac{Precision*Recall}{Precision+Recall}$$

Accuracy allows to have the fraction of the correct classifications. To calculate it, the sum of the true positive and negatives is added, divided all the samples evaluated (true positive, false negative, false positive and true negative). The formula is as follow:

$$Accuracy = \frac{tp+tn}{tp+fn+fp+tn}$$

The *tn* value in the formula indicates the number of true negatives.

In Table 8 the results obtained are shown.

In Table 8 it is possible to visualize a Precision with a value of 1, demonstrating that the method is able to correctly classify all collusive applications, avoiding wrong classifications with malware that does not perform collusive actions and trusted applications.

5 Related Work

Common antimalware are able to analyze only one sample at a time, without considering the presence of communication channels between applications. In this regard, the research community is working with the aim of developing methods for identifying channels aimed at Inter-Component Communication.

Covert channels can be classified as a means of launching a colluding attack. The authors of [27] worked on the development of a multichannel communication mechanism, called the *Multichannel Communication System* (MSYM), which allows the transfer of sensitive data securely on mobile devices. Its operation is based on the use of the VpnService[10] interface provided by Android, and is able to intercept the data of network sent, dividing them into several parts that will be mixed and encrypted through multiple transmission channels.

The accelerometer sensor contained in all Android devices represents a type of hidden channel capable of generating signals that reflect the movements of users. The data collected by the sensor can be read by malicious applications; with proper use of the vibration engine of the device it is possible to encrypt the stolen data and thus an application can influence the acceleration data to have it received and decrypted by another application. Two Android applications have been developed (representing respectively the source and the sink) [1] tested on three different smartphone models to verify this type of communication.

Magnetic sensors placed in devices such as smartphones and tablets called magnetometers, are used to calculate positioning and orientation. Attackers can exploit them to obtain information from isolated, non-networked computers. *Magneto* [18] is a type of secret communication that works by exploiting the magnetic fields generated by the CPU and can be used on air-gapped systems and between nearby smartphones. Magnetic waves can be generated by computers, changing the CPU workload; the only problem is that this covert channel only works for short distances and with low transmission speed, in fact, it is effective in unconstrained environments where wireless communications are blocked.

Among the extensions of the Android operating system we find *TaintDroid* [15]: it tracks the flow of privacy-sensitive data, which passes through third-party applications. Most of the applications downloaded by third parties are not reliable, for this reason the approach monitors in real time the access of the applications to the personal data of the users, going to see how they manipulate them.

Speaking of covert channels, we can say that they are divided into covert storage channel and covert timing channel [26]. To better understand the differences:

- the *covert storage channel* works by hiding the secret information in the data transmission protocol, thus realizing the secret communication, using payload or non-payload fields;
- the *covert timing channel* works by encoding the secret information, thus realizing the secret communication, using the time interval between packets.

---

[10] https://developer.android.com/reference/android/net/VpnService.

There are several methods for detecting specific covert timing channels: the authors of [19] performed filtering, grouping, and other operations on the communication packets. In addition, they chose a machine learning algorithm to train the classification model and achieve good detection performance.

IccTA [21] represents a tool based on the technique of static taint analysis, it takes care of recovering paths in which privacy and sensitive information are sent outside without users' knowledge and, consequently, without their permission. IccTA can detect paths within a single component or between multiple components. About 22 applications containing ICC-based privacy leaks were developed to test this.

Based on IccTA, researchers in [28] have developed a tool called *Amandroid*, which deals with leak detection with an ICC analysis. To perform this analysis, two components must be generated, respectively named Inter-Component Data Flow Graph (ICDFG) and Data Dependence Graph (DDG). For each component, the analysis of the flow and the dependence of the data is carried out and the exchange of communications between them is traced, returning a personalized analysis on Android applications.

DroidSafe [17] is a tool that performs the analysis of the flow of static information; reports the presence of sensitive data leaks on Android applications. For the development of DroidSafe the Soot Java Analysis framework was used, which works by analyzing one application at a time; this makes it difficult to detect a collusion attack, as it is caused by two or more applications.

Among the various methodologies for detecting communication threats between applications we find *MR-Droid* [22]. In particular he works on intent spoofing and collusion; it works by using a MapReduce-based framework to perform a compositional analysis of the application.

In [4] the authors developed *SneakLeak*, it is a new tool to detect the presence of collusion, based on model-checking. It scans multiple applications simultaneously and can identify a set of suspicious applications for possible involvement in a collusion attack. To test SneakLeak a set of Android applications belonging to the DroidBench dataset was considered, showing collusion through communication between apps.

6 Conclusion and Future Work

The spread of smartphones has allowed the creation of a new target for cybercriminals, who are committed to constantly creating new malware, in order to evade controls and attack people's sensitive data. Among the novelties we find the colluding attack, which is based on the communication between two or more malicious applications.

In this chapter, we proposed a method based on the use of formal methods for collusion detection. An automaton is generated for each class of an Android application, transforming the instructions of the java bytecode into CCS processes. The model checking technique is exploited to verify if the automata satisfy the prefiltering properties, so as to obtain the set of possible candidate applications to compete in a collusive one. The τ-automata of the candidate applications are then generated, which contain the push actions useful for identifying the different types of collusive attack, such as: (*SharedPreferences*, *ExternalStorage*, *BroadcastReceiver*); this is to understand if we are in the presence of applications that are able to perform a colluding action.

To then explain the rationale behind the detection of colluding applications, we propose an algorithm for explainability. In this way, it is possible to indicate whether an application is able to perform a collusive action and automatically explain why the application under analysis is able to perform the specific malicious action.

We considered several datasets with a total of 1072 applications in order to test the performance of our approach, divided as follows: ACE dataset (482 applications, all collusive), DroidBench 2.0 (120 applications, no collusion), Drebin repository of malware (100 non-colluding malware), [9] repository (20 apps, all collusive), AMD repository (50 ransomware samples, no collusion) and a set of legitimate apps from the Google Play Store (300 apps, no collusion).

As a future work, the method could be evaluated, extending the set of collusive attacks, so as to further evaluate the effectiveness of the proposed method in correctly detecting new collusive attacks. The idea is to develop a framework aimed at automatically injecting malicious payloads colluded into legitimate Android applications, thus extending the functionality of the tool proposed by the authors in [5]. Another idea is represented by the development of a method aimed to reduce the actions of the automata, with the aim of applying the formal equivalence check for collusive detection.

References

1. Al-Haiqi A, Ismail M, Nordin R (2014) A new sensors-based covert channel on android. Sci World J 2014
2. Arp D, Spreitzenbarth M, Hubner M, Gascon H, Rieck K, Siemens C (2014) Drebin: effective and explainable detection of android malware in your pocket. In: NDSS, vol 14, pp 23–26
3. Bacci A, Bartoli A, Martinelli F, Medvet E, Mercaldo F, Visaggio CA (2018) Impact of code obfuscation on android malware detection based on static and dynamic analysis. In: ICISSP, pp 379–385
4. Bhandari S, Herbreteau F, Laxmi V, Zemmari A, Roop PS, Gaur MS (2017) Sneakleak: detecting multipartite leakage paths in android apps. In: 2017 IEEE Trustcom/BigDataSE/ICESS. IEEE, pp 285–292
5. Blasco J, Chen TM (2018) Automated generation of colluding apps for experimental research. J Comput Virol Hacking Tech 14(2):127–138
6. Canfora G, De Lorenzo A, Medvet E, Mercaldo F, Visaggio CA (2015) Effectiveness of opcode ngrams for detection of multi family android malware. In: 2015 10th international conference on availability, reliability and security. IEEE, pp 333–340
7. Canfora G, Martinelli F, Mercaldo F, Nardone V, Santone A, Visaggio CA (2018) Leila: formal tool for identifying mobile malicious behaviour. IEEE Trans Softw Eng 45(12):1230–1252

8. Casolare R, Martinelli F, Mercaldo F, Santone A (2019) A model checking based proposal for mobile colluding attack detection. In: 2019 IEEE international conference on big data (big data). IEEE, pp 5998–6000
9. Casolare R, Martinelli F, Mercaldo F, Santone A (2020) Android collusion: detecting malicious applications inter-communication through shared preferences. Information 11(6):304
10. Casolare R, Martinelli F, Mercaldo F, Santone A (2020) Detecting colluding inter-app communication in mobile environment. Appl Sci 10(23):8351
11. Cimino MG, De Francesco N, Mercaldo F, Santone A, Vaglini G (2020) Model checking for malicious family detection and phylogenetic analysis in mobile environment. Comput Secur 90:101691
12. Cimitile A, Martinelli F, Mercaldo F, Nardone V, Santone A (2017) Formal methods meet mobile code obfuscation identification of code reordering technique. In: 2017 IEEE 26th international conference on enabling technologies: infrastructure for collaborative enterprises (WETICE). IEEE, pp 263–268
13. Cimitile A, Mercaldo F, Nardone V, Santone A, Visaggio CA (2018) Talos: no more ransomware victims with formal methods. Int J Inform Secur 17(6):719–738
14. Enck W (2011) Defending users against smartphone apps: techniques and future directions. In: International conference on information systems security. Springer, pp 49–70
15. Enck W, Gilbert P, Han S, Tendulkar V, Chun BG, Cox LP, Jung J, McDaniel P, Sheth AN (2014) Taintdroid: an information-flow tracking system for realtime privacy monitoring on smartphones. ACM Trans Comput Syst (TOCS) 32(2):1–29
16. Ferrante A, Medvet E, Mercaldo F, Milosevic J, Visaggio CA (2016) Spotting the malicious moment: characterizing malware behavior using dynamic features. In: 2016 11th international conference on availability, reliability and security (ARES). IEEE, pp 372–381
17. Gordon MI, Kim D, Perkins JH, Gilham L, Nguyen N, Rinard MC (2015) Information flow analysis of android applications in droidsafe. In: NDSS, vol 15, p 110
18. Guri M (2020) Magneto: covert channel between air-gapped systems and nearby smartphones via CPU-generated magnetic fields. Future Gener Comput Syst
19. Han J, Huang C, Shi F, Liu J (2020) Covert timing channel detection method based on time interval and payload length analysis. Comput Secur 97:101952
20. Iadarola G, Martinelli F, Mercaldo F, Santone A (2019) Formal methods for android banking malware analysis and detection. In: 2019 sixth international conference on internet of things: systems, management and security (IOTSMS). IEEE, pp 331–336
21. Li L, Bartel A, Bissyandé TF, Klein J, Le Traon Y, Arzt S, Rasthofer S, Bodden E, Octeau D, McDaniel P (2015) ICCTA: detecting inter-component privacy leaks in android apps. In: 2015 IEEE/ACM 37th IEEE international conference on software engineering, vol 1. IEEE, pp 280–291
22. Liu F, Cai H, Wang G, Yao D, Elish KO, Ryder BG (2017) MR-Droid: a scalable and prioritized analysis of inter-app communication risks. In: 2017 IEEE security and privacy workshops (SPW). IEEE, pp 189–198
23. Marforio C, Ritzdorf H, Francillon A, Capkun S (2012) Analysis of the communication between colluding applications on modern smartphones. In: Proceedings of the 28th annual computer security applications conference, pp 51–60
24. Mercaldo F, Visaggio CA, Canfora G, Cimitile A (2016) Mobile malware detection in the real world. In: 2016 IEEE/ACM 38th international conference on software engineering companion (ICSE-C). IEEE, pp 744–746
25. Nguyen T, McDonald J, Glisson W, Andel T (2020) Detecting repackaged android applications using perceptual hashing. In: Proceedings of the 53rd Hawaii international conference on system sciences
26. Shrestha PL, Hempel M, Rezaei F, Sharif H (2015) A support vector machine-based framework for detection of covert timing channels. IEEE Trans Dependable Secure Comput 13(2):274–283
27. Wang W, Tian D, Meng W, Jia X, Zhao R, Ma R (2020) MSVM: a multichannel communication system for android devices. Comput Netw 168:107024

28. Wei F, Roy S, Ou X (2018) Amandroid: a precise and general inter-component data flow analysis framework for security vetting of android apps. ACM Trans Privacy Secur (TOPS) 21(3):1–32
29. Xu K, Li Y, Deng RH (2016) Iccdetector: ICC-based malware detection on android. IEEE Trans Inf Forensics Secur 11(6):1252–1264

Towards Collaborative Security Approaches Based on the European Digital Sovereignty Ecosystem

Amjad Ibrahim and Theo Dimitrakos

Abstract The need for collaboration and digital transformation is among the lessons the world has realized during the pandemic. This chapter argues that the strategy of the European Union to bring the concepts of sovereignty to the digital world is a crucial enabler to achieve these two goals. This strategy is being shaped by several community initiatives, laws, and projects that tackle different aspects of digital sovereignty. In this chapter, we survey the relevant literature to present an understanding of the emerging technology and highlight a research agenda based on challenges and gaps that we identify in different topics. Next, we discuss the requirements and challenges of *identity and trust, sovereign data exchange, federated catalogues, and compliance* as digital sovereignty pillars. This discussion is helpful to researchers in identifying relevant problems and practitioners in designing future-proof solutions. Finally, we illustrate the benefits of digital sovereignty through a use-case from the domain of collaborative security approaches.

1 Introduction

The ongoing coronavirus crisis has shown the importance of digital transformation and collaboration. Information and cyber-physical systems are becoming indispensable in our societal and industrial activities. Their complexity is increasing (e.g., cloud, network), their boundaries are not fixed (e.g., IoT), and they utilize complex black-box-based components (e.g., AI). However, they must be able to collaborate in order to cope with modern challenges such as international supply-chain issues, cross-border contact tracing [1], and global vaccine certificates [2].

The European Union (EU) is leveraging these needs to propose a framework for regulating and motivating collaborations in the digital world based on the core values

A. Ibrahim (✉) · T. Dimitrakos
German Research Center, Huawei Technologies Düsseldorf GmbH, Munich, Germany
e-mail: amjad.ibrahim@huawei.com

T. Dimitrakos
School of Computing, University of Kent, Canterbury, UK

T. Dimitrakos et al. (eds.), *Collaborative Approaches for Cyber Security in Cyber-Physical Systems*, Advanced Sciences and Technologies for Security Applications,
https://doi.org/10.1007/978-3-031-16088-2_6

of its member states. *Digital sovereignty* is often mentioned as a core value that the EU aims to establish in the coming years [3–5]. This is important for the EU because of the existence of non-EU actors (mainly from the US and China) who may not necessarily, share the same values.[1] However, as we argue in this chapter, digital sovereignty is a crucial enabler, in general, for collaborative approaches, especially for security purposes.

Borrowing the political definition, we define *sovereignty* as the "supreme authority within a territory" [6]. In a digital context, authority roughly refers to the *self-determination, transparency, and control* by an individual, member state, or the EU of their digital assets (e.g., data) [3, 4]. Given the emerging technologies in networking, cloud, IoT, and AI, it becomes a challenge to define digital assets comprehensively and ensure sovereignty over them. Thus, several initiatives, communities, standards, legislation are emerging to compose the digital sovereignty ecosystem; public-private, academia-industry collaborations that tackle different aspects of sovereignty are mainly shaping this ecosystem [7]. To that end, this chapter discusses the definition, advantages, cultural understanding, perspectives, problems, opportunities, and gaps that relate to the digital sovereignty ecosystem.

Stakeholders possess distinct perspectives on the concept. For instance, market regulators (e.g., European Commission or Securities and Exchange Commission in the US) decompose the concept to a *data framework* and infrastructure, a *trustworthy environment* that considers, e.g., the EU-wide certification scheme, and a *competition and regulation framework* for protecting the potential of start-ups [3, 4]. Instead, we adopt a technical perspective that focuses on sovereignty over assets such as identity, data and information, software, workload, and operations. Specifically, we consider the infrastructure that contributes to data sovereignty in Europe with details of how this infrastructure is augmented with domain-specific tools to be operationalized in application fields (e.g., data space enablement in manufacturing).

Since data exchange is a fundamental ingredient to establish collaborations, we argue that the pillars of digital sovereignty (e.g., sovereign data exchange and trust) are an effective method to define collaborative approaches. However, as we shall see in this chapter, the digital sovereignty ecosystem comprises several rapidly evolving layers. As a result, it is challenging to understand the strategy and direction revolving around such an interdisciplinary top-level concept. Failing to establish such an understanding by cyber-security researchers and practitioners could lead to missing out on the opportunity to leverage sovereignty. For example, the reference architecture of the International Dataspaces Association (IDSA) can be used to create an interoperable sovereign cyber-security dataspace [7–9]; as such, labs, companies, and individuals can exchange their data while guaranteeing the self-determination of data owners.

To the best of our knowledge, the literature lacks a comprehensive survey of the different components of the digital sovereignty ecosystem. Such a survey of the relevant initiatives, legislation, and standards contributes to describing a research

[1] Example would be the GDPR as this set of legal norms is binding for all EU-member states but includes principles different from similar laws in the US or China.

agenda of gaps and challenges. This agenda helps researchers identify opportunities, propose novel collaborative approaches, and assist personnel on different levels to unravel challenges and opportunities revolving around digital sovereignty. This chapter intends to fill this gap and breaks down the concept, of digital sovereignty, into a set of pillars as defined by the public-private initiatives leading the effort (e.g., Gaia-X [7, 10]). This organization results from a thorough survey and requirements elicitation effort of the relevant regulations and technical documentations. Consequently, we identify several preliminary gaps and challenges in the sovereignty roadmap.

In addition, the survey and roadmap can expand the open innovation ecosystem in the mission of shaping the new technology. We also focus on compliance issues with sovereignty regulations that may arise in the future. The remainder of this chapter is structured as follows: an overview of the digital sovereignty concept and its related regulations is given in Sect. 2. Then, Sect. 3 provides a thorough perspective into the different pillars of sovereignty, their requirements, and hints at potential gaps. Then, in Sect. 4, we consolidate the pillars into a roadmap and illustrate their benefits in a use case in Sect. 5. Lastly, we conclude the chapter in Sect. 6.

2 An Overview of Digital Sovereignty

This section informs the readers about the context of digital sovereignty. Section 2.1 summarizes the relevant legislation that contribute to the concept, and Sect. 2.2 describes the overall picture of the ecosystem.

2.1 Laws and Strategy Towards Data Sovereignty

The EU has been actively pursuing its digital strategy for the past years [3]. Aiming to be a leading actor in establishing global standards, it has introduced several legislation to govern data. The core values of the EU drive this strategy, such as *data belonging to an individual, self-determination of own digital assets,*[2] *and privacy being a fundamental right*. The EU wants to ensure that the development and deployment of technology respect European values and consumer rights. The EU aims to empower its citizens to control their online and offline assets on an individual level.

On a corporate level, the EU is keen on maintaining a fair and competitive single digital market respecting its consumers' rights (e.g., data minimization and data protection). This entails thwarting foreign interference, hybrid threats to democracy, or market dominance by any corporation. Thus, the EU focuses on promoting the

[2] Historically, these core values arise from a fundamental human right developed in Germany in the 1980s called "Informationelle Selbstbestimmung", which roughly translates to informational self-determination.

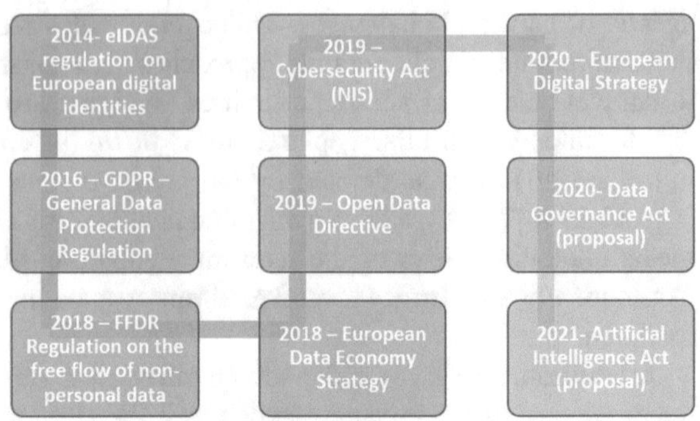

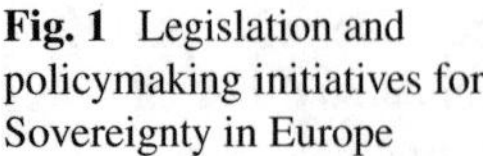

Fig. 1 Legislation and policymaking initiatives for Sovereignty in Europe

single market, in which data stays in Europe; the orderly flow of personal and non-personal data is allowed within the EU. However, the cross-border flow of non-personal data must be controlled.

As a legislative entity, the EU introduced several laws, as illustrated in Fig. 1, to ensure compliance with its values. In 2016, the General Data Protection Regulation (GDPR) was enforced to ensure the rights of a natural person to control the processing and storage of one's personal data. In 2018, the regulation on the *free flow of non-personal data* was announced. It focused on regularizing the flow, collection, and storage of anonymized data within the EU in compliance with the GDPR. In 2019, the *cybersecurity act* was added to establish an EU-wide cybersecurity certification framework for digital products, services, and processes. Also, in 2019, the *open data directive* was promoted as a foundation to create high-value data sets. The directive focused on regularizing the licensing of open data, which encouraged the reuse of public sector information.

With the laws mentioned earlier, the EU focused on the data level. However, to achieve the remaining goals of its digital transformation strategy, the EU aspires to leverage collaborative technological innovation on the infrastructure level. Specifically, the EU aims to build a trusted data infrastructure (with emphasis on integrity and resilience) on the foundation of a secure telecommunications infrastructure. This technology should enable a trusted connection between different data ecosystems, ensuring clear data access and usage specifications. It also has to be compatible with different digital identities, allowing interoperability and autonomy.

The fulfilment of the above requirements contributes to building the EU digital sovereignty infrastructure. On top of that, sovereignty will be augmented by higher-level layers that tailor to domain-specific requirements, e.g., mobility services. In that direction, the EU is also preparing specific regulations such as the Digital Services Act for Content regulation, the Data Governance Act for Industrial Data, and the AI Ethics for Personal Data.

2.2 The Digital Sovereignty Ecosystem

As introduced previously, laws provide the starting point to elicit the requirements for digital sovereignty. Now, we focus on transforming the requirements into technical concepts and introducing the architecture of digital sovereignty. First, we want to remind the reader that digital sovereignty is mainly concerned with the self-determination of digital assets.[3] In the modern information and communication technology spectrum, assets revolve roughly around four aspects: *identity, data and information, software, and operations*.[4] Firstly, *identity (self-)sovereignty* concerns protecting user identity and putting the user in control over when to whom, and how they assert their identity and how they convey their entitlements [11]. This entails removing the knowledge and power from any single commercial (or institutional) identity provider. There are ongoing European efforts in this direction, e.g., the European self-sovereign identity framework (ESSIF) [12].

Secondly, *data and information sovereignty* refer to the transparency and control over the protection, residence, and use of user data. It gives individuals, states, or the EU control over the encryption, privacy, access to data, how and why this data is processed and utilized (including monetized) by actors in a value chain. Thirdly, *software workload sovereignty* ensures that applications and workloads run securely without dependence on a specific provider software. This requires the transparency of the usage of assets used in the cloud; similarly, the control over workloads processing assets to generate value. Finally, *operational sovereignty* concerns the visibility and control over deployment paradigms, such as cloud systems and their corresponding operations.

As it is decomposed above, the requirements of digital sovereignty are being shaped by several community initiatives that tackle different aspects of sovereignty. These collaborations aim to develop a standardized infrastructure modelled to facilitate data flow and then operationalize it within data spaces [7, 8, 10]. As such, the digital sovereignty ecosystem can be described as a layered infrastructure shown in Fig. 2a, and the corresponding communities deriving them in Fig. 2b. Several projects are working on envisioning the infrastructure that enables sovereignty. For instance, standardized cloud infrastructure is targeted within the Gaia-X [10] project[5] to provide the essential functions. Also, the foundation of identity as an element of sovereignty is established within the European Self-Sovereign Identity Framework (ESSIF) [12]. The ongoing effort in this layer is mainly concerned with stan-

---

[3] While different perspectives on sovereignty may further categorize it into different classes, e.g., technological sovereignty or data sovereignty; we discuss data sovereignty in this chapter.

[4] Tech companies are also acknowledging some of these aspects, e.g., Google (see https://cloud.google.com/blog/products/identity-security/how-google-cloud-is-addressing-data-sovereignty-in-europe-2020).

[5] Several representatives of business, science, and administration launched Gaia-X, a project to create a federated data infrastructure for Europe based on values such as sovereignty, openness, transparency, innovation, and connectivity [9, 10].

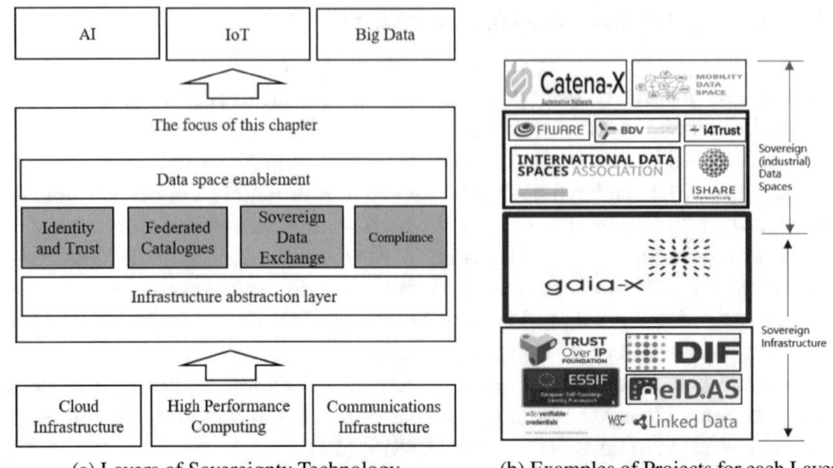

(a) Layers of Sovereignty Technology (b) Examples of Projects for each Layer

Fig. 2 An overview of digital sovereignty and its projects

dardizing the common foundations that can then be used to build sovereign systems; we elaborate on these functions in the following sections.

An enablement layer (as shown in Fig. 2a) is envisioned on top of this sovereign infrastructure to operationalize the above foundations. This layer supports sovereignty within domain-specific ecosystems that are called data spaces [7, 8]. Data spaces are built to enable unified and friendly interfaces for services that depend on data, such as AI, IoT, and Big Data for specific industries. The interfaces facilitate data flow and promote data value creation. In this layer, a set of collaborations are undergoing. As shown in Fig. 2b, many projects leverage the sovereign infrastructure provided by, e.g., Gaia-X to build sovereign data spaces. One example is the work done by the IDSA and its (more than 110) members from industry and academia [8]. Among them are companies such as Deutsche Telekom and Volkswagen and research institutes like Fraunhofer. They provided a reference framework (IDS-RAM) to facilitate the secure exchange of industrial data and services within the concept of data spaces. IDS-RAM is published as a German standard (DIN SPEC 27070) [13]. More domain-specific examples of data spaces are iSHARE and i4Trust [14, 15]. iSHARE is a collaborative effort to improve data exchange among organizations in the Dutch logistics sector [15]. i4Trust is another collaborative initiative boosting the development of innovative services around new data value chains in multiple sectors [14].

Both Gaia-X and IDSA share the goal of creating an ecosystem of trust for data sharing but focus on different layers. Gaia-X architecture and IDS-RAM can be implemented independently of each other but also be combined in a more comprehensive data sovereignty framework that covers digital sovereignty requirements in both cloud data infrastructure and data sharing and exchange layers. A position paper already proposed a conceptual mapping between IDS RAM and Gaia-X architecture [9].

2.2.1 Cloud Providers in Gaia-X

In addition to the community work, cloud providers who are members of the sovereignty ecosystem are also preparing to support sovereignty. We provide a non-exhaustive list of these providers' engagement.

- Amazon Web Services (AWS) is contributing to the initiatives of the European Commission under the Free Flow of non-personal data Regulation; an example is the code of conduct on switching cloud providers and porting data (SWIPO). They are also building technology, such as provable security (CodeGuru, IAM Access Analyzer, virtual private cloud reachability analyzer, Network reachability inspector) [16].
- Microsoft is contributing through several sovereign Azure features. For instance, the azure active directory implements verifiable credentials. The management of data residency using mechanisms that facilitate sovereignty, such as hybrid connectivity using VPN, express routes, data gateways, and supporting Azure policies. Azure policies allow creating, assigning, and managing policies governing resources. The governance concerns compliance with corporate standards and service level agreements [17].
- Google explains its engagement by claiming to provide data, operational, and software sovereignty solutions. It aims to provide customers with a "mechanism to prevent the provider from accessing their data," " provide them with assurances that the people working at a cloud provider cannot compromise customer workloads," and "provides customers with assurances that they can control the availability of their workloads and run them without being dependent on or locked-in to a single cloud provider" [18].

While this chapter focuses on the cloud and data infrastructure layer, in addition to the enablement layer, it is worth mentioning that there are other relevant aspects and challenges of sovereignty. Among these aspects is the network and communication level and how to abstract it so that sovereign infrastructure and data spaces operate seamlessly. Consequently, existing infrastructures will risk being replaced if they do not comply with the laws and regulations targeting vendor lock-in. Another aspect of digital sovereignty's big picture lies within the emerging data ecosystems enabled by sovereign technology. There, challenges related to trustworthy AI will arise. Specifically, questions regarding fairness, privacy, responsibility, accountability, explainability, and transparency would appear within these distributed computation platforms. Also, privacy-related challenges, e.g., secure multi-party computation, homomorphic encryption, private computing, differential privacy, and federated learning, would arise within sovereignty.

3 Pillars of Sovereignty

This section discusses the main technical concepts that are building-up sovereignty. We start in Sect. 3.1 with the *identity* aspect of digital sovereignty, which corresponds to the identity and trust component within Gaia-X [10] and ESSIF [12]. We begin with a brief recap of the concept and discuss its functionality, challenges, and gaps. Section 3.2 discusses the second pillar, which is a sovereign fabric for data exchange, i.e., a technology to share and control usage of data securely. Similar to the previous section, we present the concept and decompose it into high-level requirements as discussed within initiatives such as Gaia-X and IDSA, and we conclude with potential gaps. To support the services of identity management and sovereign data exchange, the technical specifications of digital sovereignty (e.g., Gaia-X technical documentation) describe two essential services within the federated services: compliance and federated catalogues. In Sects. 3.3 and 3.4, we consider them.

3.1 Identity (Self-)Sovereignty

The world has historically split on (digital) identity based on different cultural perspectives. For instance, while the EU member states endorse institutionalized national and cross-border digital identities, to some extent, Anglo-American societies tend to reject public identity infrastructures. Moreover, some nations tend to distrust decentralized private sector processing of their identity data, while others distrust central government access to sensitive identity information. This discrepancy has found its way into the digital world through different trust models or paradigms (federated or decentralized) of Identity and Access Management (IAM) systems [11, 19]. Digital sovereignty aims to bring sovereignty to identity regardless of the used paradigm or trust framework.

As mentioned, identity sovereignty gives users control over when to whom, how they assert their identity, and how they convey their entitlements. This coined the concept of Self-Sovereign Identity (SSI) [11, 19]—a model to enable users to create and control their identity, eliminating any authority managing a central infrastructure for everyone. SSI is based on Decentralized Identifiers (DIDs) that support claims about the identity, e.g., the identity holder is fully vaccinated against Covid-19 (the reader may refer to these papers [11, 19] for details). On a European level, an initiative to implement a self-sovereign identity (SSI) feature is planned within ESSIF [12]. Because of its distributed nature, SSI implementations raise several GDPR compliance concerns [20].

3.1.1 Requirements

While decentralization of identities seems suitable for sovereignty, data sovereignty requirements, e.g., from Gaia-X or IDSA, still support federated identity systems [21]. Specifically, the Gaia-X IAM framework supports two different approaches (participants choose which one to use): "Federated Identity Approach (e.g., OpenID Connect OIDC)" or "Decentralized Identity Approach (e.g., DID/VC)." This is reasonable because Gaia-X aims to connect centralized and decentralized infrastructures creating a trustworthy data-exchange environment. Furthermore, supporting legacy IAMs keeps the door open for more platforms while maintaining the sovereignty of users. This overall requirement sets the challenge in this domain, i.e., enabling sovereign identity even in the existence of different trust frameworks.

We elaborate on the high-level functional requirements to achieve the aforementioned goal. We mainly borrow the composition presented in the Gaia-X IAM technical documents and the Gaia-X Federation Services (GXFS) documents.[6] We highlight the crucial requirements that must be carried out by any sovereign identity and access management system.

1. Holistic IAM: covering essential IAM functions such as authentication and authorization, credentials management and verification, and decentralized identity management. This includes the dynamic mapping of credentials to permissions [21].
2. The management of organization credentials: enabling a participant to interact with the SSI-based ecosystem in a trustworthy and secure manner. This can be achieved by utilising verifiable presentations to establish policy enforcement as the base for trustful information.
3. The management of personal credentials: providing a wallet for the user.
    a. Enablement of users (holders) to interact technically with the DID-based ecosystem in a privacy-preserving way.
    b. Supporting user representation by enabling them to hold the acquired distributed identities while providing ways to disclose certain attributes for authentication and service consumption selectively.
4. Policy foundation for trust services: policy-based functionality to ensure a notion of trust as agreed on in policies [22]. As such, providers can present their issued attestations and certificates to the consumer to decide based on their policies and rules.
5. Decoupling Trust Frameworks: the technical implementation does not depend on one trust framework, e.g., electronic Identification and Trust Services (eIDAS) [23, 24]. Instead, a verifier can integrate new trust domains and verify transactions automatically.

[6] GXFS represent the minimum technical requirements and services necessary to operate the federated Gaia-X Ecosystem of infrastructure and data.

3.1.2 Challenges, and Gaps

Based on our requirements analysis, we see the following **gaps** and open challenges.

Challenge 1—Federated identity decentralization: to the best of our knowledge, no previous solution provides an easy solution that bridges federated with decentralized identity for interoperability. Thus, supporting both legacy federations such as OIDC and emerging decentralized identities such as SSI (and ESSIF) is a significant challenge. A solution to this challenge should provide a semantically correct transformation of credentials and protocols into a decentralized identity fabric.

Challenge 2—Situation-aware and privacy-preserving identity trust management: we argue that ESSIF is still missing an advanced policy-based trust management capability that integrates different trust frameworks. Currently, most initiatives are trying to integrate SSI with only one trust framework, i.e., eIDAS [23]. eIDAS is a regulation that governs the trust domain of national legal electronic identification in the European context; its relevance to the private sector has been limited. Thus, we anticipate that other SSI systems based on different frameworks will emerge. Then, being able to issue, present and verify credentials based on trust frameworks is the second challenge for self-sovereign identity. A solution in this domain should support a context-dependent use of anonymous credentials and disclosure of attributes (a preliminary solution is proposed in [24]).

Challenge 3—Trustworthy credentials retention and utilization: issues of trust impose a significant challenge for ESSIF. Instilling confidence in online interactions among relevant stakeholders is essential for the adoption of SSI. However, methods to guarantee a presented identity and, thus, respective VCs belonging to the claimed entity are still missing. Typically, owners control their identities by possessing software-based cryptographic keys associated with their entities. Such a scheme raises concerns originating from how keys are managed. The solution here should aim for a strong guarantee of binding the identity to its owner.

3.2 Sovereign Data Exchange Fabric

Sovereign data exchange enables individuals, states, or the EU to control their data's protection, privacy, access, and use—how and why it was processed, shared, and monetized by actors in a value chain. Once provided, this ability fosters innovations through novel ecosystems and data spaces. In cybersecurity research, also within Gaia-X and IDSA, such an ability is associated with transparency and *control of data usage*. Usage Control (UCON) [25–28] is a crucial enabler for data sovereignty because it allows parties to monitor resource access continuously and, potentially, revoke its usage in case of policy violation; thereby implementing technical measures to adhere to legal obligations. Further to access control, UCON is concerned with requirements that pertain to future data usage (i.e., obligations). If the resource is data, then we are concerned with data-centric UCON. Sovereignty documentation mentions data-centric UCON as a key technology [21]. It emphasizes the importance of policy enforcement because it guarantees that restrictions and obligations are realized even after access to data has been granted.[7]

To the best of our knowledge, there is no high quality, robust, efficient, and open-source implementation of UCON. Several non-profit associations are considering the same problem; examples include IDSA [8], Gaia-X European Association for Data Cloud, AISBL (Gaia-X AISBL), Catena-X [29], and iShare [15]. However, the outputs of these initiatives are mostly reference architectures and specifications. We mention several gaps that still need to be analyzed and addressed before building an efficient and comprehensive data-centric UCON. In addition, we consider the extension of this technology to include data leak tracing and recovery ability, privacy-enhanced computations, and provable conformance of sharing agreements.

3.2.1 Requirements

Let us look into more details on the requirements of sovereign data exchange. Then, we summarize and highlight key Gaia-X (and GXFS), IDSA, and iShare requirements regarding policy and usage control.

1. Data Sovereignty Services: provide participants with the capability to be entirely self-determined regarding exchanging and sharing their data. Participants can also decide to act without having the Data Sovereignty Service involved if they wish to do so.
2. Support of Gaia-X/IDS Policies: describe invariants that must be assured in a software execution environment based on the self-descriptions of assets and participants. The policies are dynamically evaluated at runtime, during onboarding and instantiation.
3. Data-centric Usage Control: includes the ability to specify policies in machine-readable, human-readable, and interoperable ways. The policies express technical, organizational, and legal conditions.
   a. Administration of Usage Policies (creation and maintenance)
   b. Enforcement of Usage Policies:
      i. Different functions for different phases of data lifecycle e.g., before, during, and after a transaction
      ii. Different notions of obligations require domain-specific mechanisms for enforcement
      iii. Auditability for transparency
      iv. Organizational measures to enforce usage policies must also be considered, as well as legal measures.

4. Policy-Driven Workload Control: restrictions confirm the mobility of service instances. For example, specific tasks must be performed by service instances from providers with a defined certification level.

[7] "While access control restricts access to specific resources (e.g., a Service or a file), data sovereignty is additionally supported with Data-Centric Usage Control" [21].

5. Data leak tracing and recovery: if some participants violate deployed usage policy, and, e.g., leak the accessed data, these participants should be traced.

3.2.2 Challenges, and Gaps

We identified the following challenges according to our elicitation and analysis of requirements related to sovereignty.

Challenge 1—Building a usage control engine with a multi-administrative delegation of authority: enabling individuals to create policies while recognizing the authority that allows them to do that for a limited duration and delegate specific capabilities to others [25]. The crucial specifications are:

- Confidentiality and integrity of data objects (attributes, policy, request)
- Verification and validation of data objects issuers
- Rigorous specification for policies with an automatic verification method that ensures the correct implementation of UCON
- Automated analysis methods to assist writing and maintaining policies.

Gap: no existing solution that delivers good performance while ensuring policy security.

Challenge 2—Data-centric usage control with automated orchestration of auditable, contextualized enforcing of obligations: extending the policy engine (from Challenge 1) with a specific enforcement layer that supports automated data-related obligation enforcement, e.g., "delete data and all its transformations after three days," or "anonymize the data before usage." The following specifications are important to tackle this challenge:

- Control of data lifecycle: tailoring usage control models to the data lifecycle phases, e.g., collection, storage, and destruction.
- Data Protection & Privacy related policies: making the ecosystem of UCON aware of protection concepts, e.g., defining policies to capture user consent.

Gap: no existing efficient solutions that support data-centric policies with self-adaptive automation.

Challenge 3—Advanced Privacy-preserving Sovereign Policy Enforcement

- Utilize privacy-protection techniques such as multi-party computation, or differential privacy to allow value to be extracted from data while meeting compliance requirements on untrusted environments.
- Automated privacy-enhancing or preserving obligations enforcement, e.g., data column masking, row filtering, anonymization, and user notifications.

Gap: no existing solutions fulfil the requirements above effectively.

Challenge 4—Operationalization of Data spaces [30]:

- (Technology) Introducing a data integration concept that eliminates physical integration and any shared common schema among data. In addition to data sovereignty, traceability, and trust management (which are also covered by the infrastructure requirements), we also need to consider data networking, co-existence, nesting, and overlapping. Efforts in this direction are still in early stages, e.g., Eclipse Data-space connector.

- (Business) Forming innovative concepts for data collaborations. This includes new collaboration formats, sharing the purpose of data exchange, and decision-making for data governance.
- (Business and Technology) Addressing the questions around federators roles. For instance, who is the federator in an ecosystem? What is their business model? And what are the implications of regulations on concrete instantiations of the federator?

Gap: the technology part of the above requirements lacks innovative solutions to operationalize data-spaces.

Challenge 5—Data sharing agreements:

- Provable conformance of usage control policies.
- Continuous monitoring, predictive analysis of agreement violation risk and adaptive reaction to mitigate non-conformance.
- Decentralized execution of data sharing agreements and smart contracts to automate sovereign data flow throughout multi-party workloads.

Gap: trustworthiness guarantees to assure participants about fulfilment of agreements are required.

Challenge 6—Data leak detection, tracing and recovery:

- Strong binding of ownership information and usage policy to data through advanced cryptographic techniques.
- Steganography inspired embedding of digital watermarks containing provenance information about ownership, point of exposure and enforceable data usage policy into digital content (including for example video content, neural network models).
- Providers use an obligation management system to handle information lifecycle based on preferences and organizational policies.

Gap: methods to protect the data beyond the digital medium are required in the context of sovereignty.

3.3 Compliance

The concept of digital sovereignty relies heavily on notions of trust and policy enforcement. The technical interpretation of these two concepts may raise questions about their compliance. Further, the necessity of compliance is also demanded based on legal obligations such as GDPR. Thus, a layer of compliance services should be added to augment trust and sovereign exchange services. A compliance service is a technical component that verifies that an agreement (in the form of, e.g., policy) was respected; for example, it shows provable evidence that "data was deleted after it was only used for analysis." This layer includes mechanisms to ensure a participant's adherence to the security, privacy, transparency, and interoperability rules. This is important when onboarding a new participant to the ecosystem. For instance, onboarding and accreditation workflow is a procedure described by Gaia-X to ensure and document that all participants, assets, and service offerings undergo a validation process before being added [21]. The ability to audit and prove compliance with the documented process is a crucial requirement.

Auditability and compliance require continuous automated monitoring functionality. This allows monitoring the compliance based on self-descriptions (also known as catalogues, which are discussed next). Monitoring also includes the updates to service offerings that should trigger revisions or re-certifications for compliance, monitoring the suspension of service offerings, and monitoring the revocation of service offerings.

The functionality of providing an auditable process is a goal that must be considered in each component of the sovereignty services. The evidence's generation, collection, granularity, storage, and security have to be tackled. The standardization of the evidence is a challenge since it varies according to the different services. For instance, how can the system prove the enforcement of an obligation function or a policy decision? Also, how can we ensure that the evidence is, e.g., authentic and tamperproof? If we increase transparency to allow compliance, how do we cope with this increase's privacy concerns? These are all open questions that must be tackled in sovereign systems. The concrete challenges are the following.

- Provide machine readable encoding of rules and obligations for:
  - Conducting conformance assessment of participants, resources and service offerings in a digital sovereignty enabling infrastructure (e.g. Gaia-X) or ecosystems (e.g. IDS) operating on top of such infrastructure.
  - Collecting proofs about the conduct of conformance assessment.
  - Verifying the integrity and authenticity of the evidence and checking the corresponding proofs of conformance against such evidence.
  - Causality analysis of evidence and attribution of responsibility for corrective actions.

- Define and facilitate the automation of Onboarding and Accreditation Workflows
  - Participants, Assets, Resources and Service Offerings undergo a validation process before being added to a Catalogue.
  - On-boarded Conformance Assessment Authorities (CABs) are competent to offer accredited conformance assessment services to participants.
  - Participants negotiate trust to share credentials and substantiate them with the corresponding proofs and evidence of assessment supported by recognized CABs.
  - Sharing evidence of verification of the credentials supporting each claim. Evidence attesting proofs is securely shared by CABs with authorized requestors through a hierarchy of federated verifiable data registries so that assessment proofs is automatically checked.
- Measure trust in transactions
  - Trust negotiation: Incremental and conditional disclosure of sensitive information to evidence accreditations.
  - Adaptation of data sharing, usage and protection mechanisms based on trust level.

3.4 Federated Catalogues

The last part of this section discusses the so-called federated catalogues [21]. We first need to introduce self-descriptions according to the Gaia-X terminology to understand the concept. Self-descriptions express characteristics of assets, service offerings and participants (represented technically as JSON-linked data objects). They may include claims about these characteristics that an issuer asserts. Self-descriptions are machine and human interpretable without any dependence on a specific technology. Their semantics and validation rules are sound and clear, and they can be easily extended. Self-descriptions are accompanied by statements of proof referenced from any trusted data infrastructure in a unique, decentralized fashion.

Within the architecture proposed by Gaia-X, self-descriptions are contained and composed within catalogues to offer services. The catalogues are synchronized and discoverable in index repositories to enable providers and their offerings. The catalogues follow a hierarchal layout that comprises federated catalogues (top-level catalogue operated by Gaia-X), ecosystem-specific catalogues (e.g., Catena-X), company catalogues (internal use). The use of self-descriptions organized in catalogues empowers different participants in their decision-making processes. However, as we also hinted about identity management, this is conditioned on providing a trustworthy verification mechanism to verify claims. The verifiability of claims in the existence of heterogenous identity deployments (centralized or decentralized) is the challenge here. Further, the enforcement, continuous validation, and trust monitoring of these dynamic catalogues are crucial; such requirements are an excellent candidate for tackling by usage policies. To that end, the goals scoped in this domain can be summarized as follows.

- Verifiable catalog federation that includes:
  - Linked data model with associated Verifiable Credentials (VCs).
  - Automated verification of self-descriptions leveraging linked data.
- Verifiable catalog entry life-cycle management
  - Continuously monitor linked data life-cycle, validity and verifiability.
  - Verifiable and propagate amendment of self-descriptions and associated verifiable presentations.
- Verifiable agreement initiation
  - Automated discovery, verification and auditable enforcement of catalog entries and multi-lateral agreement of smart contracts and usage policies.
  - Production of auditable evidence collection scheme for validating conformance during the enforcement of the agreement.

4 Technical Roadmap

This chapter proposes a technical roadmap for sovereignty; we focus on sovereign technology stack challenges. As a result of a thorough review of technical specifications and regulations of sovereignty and relevant concepts, we summarize a vision (as a set of preliminary gaps that can be tackled) in each pillar. Figure 3 plots the concrete recommendations for each pillar.

(A) *Identity and trust are concerned with* continuous authentication and authorization in compliance with emerging identity schemes. An essential concept of trust is Self-Sovereign Identity (SSI), enabling users to create and control their digital identities. With different identity and access management technologies, trust models, and trustworthiness concerns, enabling SSI remains an open challenge of digital sovereignty. Researchers and practitioners can tackle this challenge by building a policy-based bridge.

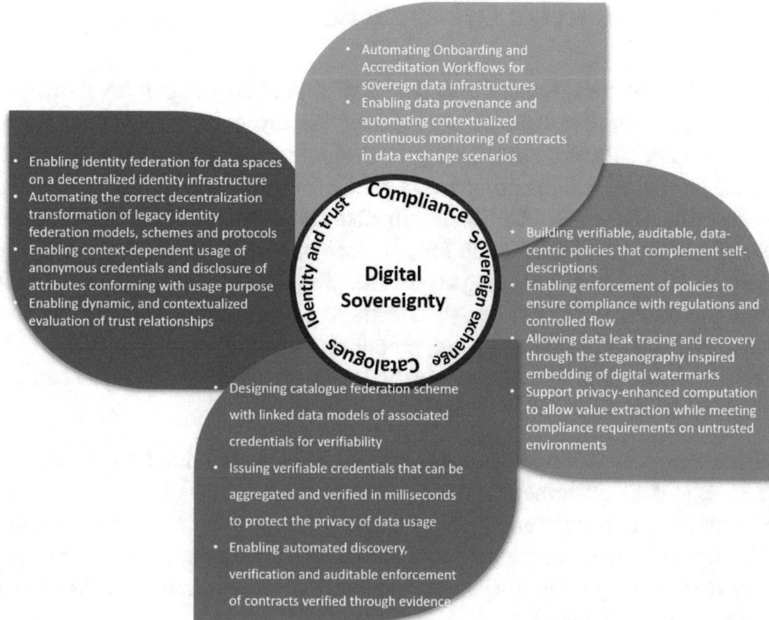

Fig. 3 The concrete items discussed in this chapter

(B) *Sovereign data exchange* allows individuals, states, or the EU to control the protection, privacy, access, use of their data. Thus, data-centric usage control systems must enforce restrictions and obligations throughout the data exchange life cycle with the ability to verify enforcement retrospectively. However, our analysis shows that robust advanced technology for usage control that caters to modern use cases and requirements is still missing. Therefore, researchers and practitioners can fill this gap by equipping usage control technologies with features like multi-administrative delegation of authority, automated orchestration of data-centric obligation enforcement, and awareness of privacy-preserving obligations.

(C) *Compliance* services augment trust and sovereign data exchange by ensuring participants' adherence to the security, privacy, transparency, and interoperability rules. For that, auditability and continuous automated monitoring are required in each part of the digital system. Therefore, we propose that researchers extend policy-based control systems with traceability and evidence-support abilities at the various points of their operations.

(D) The emerging digital ecosystems would require *catalogues (directories)* to enable engagement among participants. Therefore, a trustworthy automated mechanism to verify the participants' self-descriptions is crucial.

5 Use Case: Sovereign Cyber-intelligence Dataspace

In this section, we consolidate the concepts of digital sovereignty to showcase the use case of collaborative security approaches. We present a fictional example that illustrates the benefits of the technology in designing a sovereign cyber-intelligence data space similar to what is proposed in the C3ISP project.[8]

We describe a Sovereign Cyber-Intelligence Dataspace (SCID). In SCID, participants exchange threat information for purposes of attack detection and prevention, research, reporting, and increasing awareness [31]. The founding principle in SCID concerns protecting the rights of data owners, e.g., business companies, by enforcing constraints on the data share, e.g., obtaining the chief security officer's consent before transferring data to a research institute. The participants of SCID can be any subset of the following.

- End-customers are individuals using certain services provided by business companies, e.g., bank customers.
- Business companies are commercial entities that provide services to customers (individuals or other companies) in a particular domain, e.g., a bank. Such companies may leverage digital systems internally (used by their staff) or externally (used by their customers).
- Security companies are particular companies that operate in cyber security domains. For example, they monitor attacks, publish reports, or build security products, such as anti-virus software.
- Research institutes are non-profit organizations that study cyber security.
- National security agencies are official units that monitor cyber threats and help put forward legislation or standards in cyber defence.

The above participants produce and consume cyber-threat information. According to the National Institute of Standards and Technology (NIST), threat information includes any data item that benefits an organization to protect itself against a threat [32]. The major types of cyber-threat information include technical indicators, tactics, techniques, procedures, security alerts, threat intelligence reports, and security tool configurations [32]. For instance, indicators are evidence that illustrates attacks have occurred, e.g., network or host event logs, phishing emails, or code hashes.

Building a sovereign threat-information dataspace requires a similar infrastructure to the concepts presented in this chapter, i.e., identity and trust, sovereign data-exchange fabric, federated catalogues, and compliance services. Firstly, it is plausible that malicious actors would try to participate in SCID to acquire knowledge; thus, certifying the identity of participants and onboarding them is very critical. Secondly, it is very plausible that the participants in SCID are business competitors, and hence, they need the ability to determine with whom to share their data. Also, since threat data can also include personal data, end customers need to control who accesses the

[8] https://c3isp.eu/.

data and what purpose. Thirdly, to regulate the process of offering and searching for data, a catalogue is required. The security and certification of this catalogue and the agreements are essential, given the sensitivity of the data. Lastly, according to the nature of the dataspace, methods to assess participants' compliance with the data-sharing agreements would be crucial to raising trust.

The aforementioned requirements form the main foundation of the standards, laws, projects, and communities building the EU digital sovereignty ecosystem. As such, dataspaces for collaborative security approaches, e.g., SCID, will emerge based on sovereignty technology. The benefits of SCID can be illustrated by the following user scenarios supported with this technology.

Sharing for detection/prevention:

- As a customer, I want to send the security-related events of my online banking usage to the bank to review them only from machines within the bank IP range and within working hours.
- As a customer, I want to consent to use my data when there are no ongoing attacks, only if the usage is securely recorded.
- As a company, we want our customers' personal data to be anonymized before being shared with National security agencies.
- As a security company, we want to share our recommended tool configurations are recommendations only with non-profit research institutes.

Sharing for research:

- As a company, we want to transfer our clients' data to public research institutes, only if the data is anonymized and the location is cloaked, and any machine learning algorithm using the data is differentially private with a certain epsilon value. Also, we want each client to be notified by email about what exact study was their data used in.
- As a customer, I want to give (or deny) consent for requests by security companies to process my data on my hardware, and revoke it as I wish.
- As a security company, we want to participate in research projects only if our data is stored on secure hardware in the same city, and the data is deleted after three months.

6 Conclusion

The EU wants to ensure Europe's ability to act independently within the emerging technologies. Therefore, it introduces the concept of (European) digital sovereignty through several legislation and initiatives. Digital sovereignty is concerned with giving self-determination, transparency, and control by an individual, member state, or the EU of their assets. The emerging sovereign systems would be an effective method to design collaborative approaches that require data exchange. Thus, developing an early understanding of digital sovereignty is very important because it aids researchers to understand the scope of futuristic, relevant research problems. Also, it motivates security practitioners to adopt the practices that facilitate accommodating the emerging requirements of sovereignty. Further, the EU is using industry-academia and public-private partnerships to implement its strategy, and hence, the potential to engage in shaping the technology exists.

All that said, the digital sovereignty ecosystem builds on several initiatives, standards, laws, and communities. As a result, it is still challenging to form a comprehensive view of sovereignty. Thus, this chapter reviewed the technical specifications of the projects and laws related to sovereignty and discussed the main pillars of sovereignty in depth. This discussion helped identify gaps to fill in each domain and then propose a roadmap of sovereignty that we summarize in the following.

Identity and trust are concerned with continuous authentication and authorization in compliance with emerging identity schemes. An essential concept of trust is Self-Sovereign Identity (SSI), enabling users to create and control their digital identities. With different identity and access management technologies, trust models, and trustworthiness concerns, enabling SSI remains an open challenge of digital sovereignty. We identified the harmonization of identity schemes as a problem that needs to be solved.

Sovereign data exchange fabric allows individuals, states, or the EU to control the protection, privacy, access, use of their data. Thus, data-centric usage control systems must enforce restrictions and obligations throughout the data exchange life cycle with the ability to verify enforcement retrospectively. However, our analysis showed that robust advanced technology for usage control that caters to modern use cases and requirements is still missing. Therefore, our recommendation is to research how to equip usage control technologies with features like multi-administrative delegation of authority and awareness of privacy-preserving obligations.

Compliance services augment trust and sovereign data exchange by ensuring participants' adherence to the security, privacy, transparency, and interoperability rules. For that, auditability and continuous automated monitoring are required in each part of the digital system. Furthermore, we discussed the need to extend the policy-based control systems with traceability and evidence-support abilities at the various points of their operations. Lastly, the emerging digital ecosystems would require *catalogues (directories)* to enable engagement among participants. Therefore, a trustworthy automated mechanism to verify the participants' self-descriptions is a crucial challenge that must also be tackled to build sovereign collaborative systems.

Acknowledgements The authors would like to acknowledge the contribution of the following colleagues: Bithin Alangot, Subhajit Bandopadhyay, Isabelle Hang, Ali Hariri, Ioannis Krontiris, Athanasios Rizos, Tian Wenyuan, Teng Wu, Xuebing Zhou at Huawei's Munich research center.

References

1. Markus I, Steffen G, Lachmann R, Marquis A, Schneider T, Tomczyk S, Koppe U, Rohde AM, Schink SB, Seifried J et al (2021) Covid-19: cross-border contact tracing in Germany, February to April 2020. Eurosurveillance 26(10):2001236
2. Cascini F, Causio FA, Failla G, Melnyk A, Puleo V, Regazzi L, Ricciardi W (2021) Emerging issues from a global overview of digital covid-19 certificate initiatives. Front Pub Health 9. https://doi.org/10.3389/fpubh.2021.744356. https://www.frontiersin.org/article/10.3389/fpubh.2021.744356
3. A digital future for Europe. https://www.consilium.europa.eu/en/policies/a-digital-future-for-europe/
4. Council A The European union and the search for digital sovereignty
5. Roberts H, Cowls J, Casolari F, Morley J, Taddeo M, Floridi L (2021) Safeguarding European values with digital sovereignty: an analysis of statements and policies. Internet Policy Rev
6. Philpott D (2020) Sovereignty. In: Zalta EN (ed) The Stanford encyclopedia of philosophy, Fall, 2020th edn. Stanford University, Metaphysics Research Lab
7. Braud A, Fromentoux G, Radier B, Le Grand O (2021) The road to European digital sovereignty with GAIA-X and IDSA. IEEE Netw 35(2):4–5. https://doi.org/10.1109/MNET.2021.9387709
8. Otto B, Hompel MT, Wrobel S (2019) International data spaces. In: Digital transformation, pp 109–128. Springer
9. Otto PDB (2021) GAIA-X and IDS. https://doi.org/10.5281/zenodo.5675897
10. C. AISBL E Gaia-X: A federated and secure data infrastructure. https://www.gaia-x.eu/
11. Mühle A, Grüner A, Gayvoronskaya T, Meinel C (2018) A survey on essential components of a self-sovereign identity. Comput Sci Rev 30:80–86
12. Pastor M Essif functional scope
13. Otto BS IDS reference architecture model. Version 3.0. https://internationaldataspaces.org/publications/ids-ram/
14. i4trust Data spaces for effective and trusted data sharing. https://i4trust.org/
15. iSHARE: Sharing logistics data in a uniform, simple and controlled way. iSHARE. https://www.ishareworks.org/en
16. Peterson M What's next for Europe's data revolution? AWS joins the GAIA-X initiative. https://aws.amazon.com/blogs/publicsector/what-next-europes-data-revolution-aws-joins-gaia-x-initiative/
17. Microsoft Azure active directory verifiable credentials documentation. https://docs.microsoft.com/en-us/azure/active-directory/verifiable-credentials/
18. Kurian G Engaging in a European dialogue on customer controls and open cloud solutions. https://cloud.google.com/blog/products/identity-security/how-google-cloud-is-addressing-data-sovereignty-in-europe-2020
19. Tobin A, Reed D (2016) The inevitable rise of self-sovereign identity. Sovrin Found 29
20. Giannopoulou A (2020) Data protection compliance challenges for self-sovereign identity. In: International congress on blockchain and applications. Springer, pp 91–100
21. GAIA-X (2020) Technical architecture. https://www.data-infrastructure.eu/GAIAX/Redaktion/EN/Publications/gaia-x-technical-architecture.pdf
22. Software requirements specification for GAIA-X federation services trust services API IDM.TSA (2021). https://www.gxfs.de/federation-services/identity-trust/trust-services/
23. Dumortier J (2017) Regulation (eu) no 910/2014 on electronic identification and trust services for electronic transactions in the internal market (eidas regulation). In: EU Regulation of e-commerce. Edward Elgar Publishing
24. Kubach M, Roßnagel H (2021) A lightweight trust management infrastructure for self-sovereign identity. Open Identity Summit 2021
25. Bandopadhyay S, Dimitrakos T, Diaz Y, Hariri A, Dilshener T, La Marra A, Rosetti A (2021) Datapal: data protection and authorization lifecycle framework. In: 2021 6th South-East Europe design automation, computer engineering, computer networks and social media conference (SEEDA-CECNSM). IEEE, pp 1–8

26. Dimitrakos T, Dilshener T, Kravtsov A, La Marra A, Martinelli F, Rizos A, Rosett A, Saracino A (2020) Trust aware continuous authorization for zero trust in consumer internet of things. In: 2020 IEEE 19th international conference on trust, security and privacy in computing and communications (TrustCom). IEEE, pp 1801–1812
27. Hariri A, Bandopadhyay S, Rizos A, Dimitrakos T, Crispo B, Rajarajan M (2021) Siuv: a smart car identity management and usage control system based on verifiable credentials. In: IFIP international conference on ICT systems security and privacy protection. Springer, pp 36–50
28. Park J, Sandhu R (2002) Towards usage control models: beyond traditional access control. In: Proceedings of the seventh ACM symposium on access control models and technologies, pp 57–64
29. Catena-X Automotive network. https://catena-x.net/en/
30. Otto B (2021) Interviewee, talk at the security and trust summit
31. Tounsi W, Rais H (2018) A survey on technical threat intelligence in the age of sophisticated cyber attacks. Comput Secur 72:212–233
32. Johnson CS, Feldman L, Witte GA et al (2017) Cyber threat intelligence and information sharing

POM: A Trust-Based AHP-Like Methodology to Solve Conflict Requirements for the IoT

Davide Ferraris, Carmen Fernandez-Gago, and Javier Lopez

Abstract The Internet of Things (IoT) is an environment of interconnected entities which are identifiable, usable and controllable via the Internet. Trust is necessary for a system such as the IoT as the entities involved should know the other entities they have to interact with. In order to guarantee trust in an IoT entity, it is useful to consider it during all its System Development Life Cycle (SDLC). The requirements phase is one of the first and the most important phases of the SDLC. In this phase, trust requirements must be elicited in order to guarantee that the built entity can be trusted. However, during this phase, it is possible to raise conflicts among requirements reflecting conflicting needs. Decision-making processes can be helpful in order to solve these issues. The Analytic Hierarchy Process (AHP) is a discipline that supports decision-makers in choosing between heterogeneous and conflicting alternatives, but it has several problems, especially if there are numerous parameters. Thus, we propose an AHP-like methodology called Pairwise Ordination Method (POM). Its aim is to solve issues among conflicting requirements deciding which one is the less important in order to modify or delete it, maximising the trust level of the IoT entity under development.

1 Introduction

Trust is a difficult concept to be defined because it is strongly connected to its context [4]. Moreover, according to Pavlidis [18] and Hoffman et al. [10], trust is strongly related to other domains such as privacy, identity and security. As stated by Ferraris et al. [6], this consideration can be even more important in developing an Internet

D. Ferraris (✉) · J. Lopez
Department of Computer Science, University of Malaga, Malaga, Spain
e-mail: ferraris@lcc.uma.es

J. Lopez
e-mail: jlm@lcc.uma.es

C. Fernandez-Gago
Department of Applied Mathematics, University of Malaga, Malaga, Spain
e-mail: mcgago@lcc.uma.es

T. Dimitrakos et al. (eds.), *Collaborative Approaches for Cyber Security in Cyber-Physical Systems*, Advanced Sciences and Technologies for Security Applications,
https://doi.org/10.1007/978-3-031-16088-2_7

of Things (IoT) entity. About this topic, Ferraris et al. [7] proposed a framework for the IoT in order to develop a trusted IoT entity. In this framework, it is fundamental to consider trust holistically in the whole System Development Life Cycle (SDLC).

During the SDLC of an IoT entity, it is possible to encounter conflicts among needs, requirements or models. A useful method to solve conflicts among heterogeneous alternatives is represented by the Analytic Hierarchy Process (AHP) developed by Thomas L. Saaty [20]. AHP is one of the most implemented methodologies in the Multi-Criteria Decision Analysis (MCDA) [22], a discipline that supports the Decision-Maker (DM) in choosing among numerous and conflicting alternatives. This methodology allows the DM to compare different alternatives in order to decide which one better fulfils a specific goal considering both qualitative and quantitative criteria. At the end of the process, this methodology returns a global value for each criterion and alternative. According to this value, the DM is able to choose the best alternative among the others. However, AHP raises some issues. In fact, it is possible to have inconsistent choices. This risk increases when the number of alternatives and the criteria grow.

In order to avoid this issue, we propose an AHP-like methodology called Pairwise Ordination Method (POM). Moreover, we believe that POM is useful in an environment such as the IoT because, due to the uncertainty and the heterogeneity of this topic, we are able to decide among different alternatives belonging to various fields. AHP was developed to implement pairwise comparison, taking into account that the human brain has a limited capacity to make decisions if it has to choose among many alternatives [3]. POM also guarantees this aspect, which can be very useful also for IoT entities with limited computational power [17].

The paper is structured as follows. Section 2 describes the related work about IoT and trust and the background related to the K-Model and AHP. In Sect. 3, we explain the POM methodology, while in Sect. 4, we describe an IoT use case scenario to illustrate how to implement POM. In Sect. 5, we discuss the results. Finally, in Sect. 6, we conclude the paper and discuss future work.

2 Related Work

In this section, we present the state of the art about IoT and trust, then we present our K-Model originally developed in [7] because the decision-making process is an important transversal activity of our framework. Then, we discuss what the AHP is and how it has been implemented in the state of the art.

2.1 IoT and Trust

The IoT is a network of interconnected objects. Roman [19] states that the goal of the IoT is to enable smart entities to be connected anytime, anyplace, with anything and anyone, ideally using any network and any service. It is expected that these entities will have to interact with each other often under uncertain conditions.

Mechanisms to solve this lack of information are needed and trust can help address this need by overcoming uncertainty [5].

Trust is a difficult concept to define "because it is a multidimensional, multidisciplinary and multifaceted concept" [24].

Jøsang [11] defines trust as personal and subjective. For McKnight [15] trusting someone means to depend on him/her, no matter the consequences. Gambetta [9] recognises trust and the symmetrical distrust as a subjective probability to perform an action. Agudo et al. [1] define trust as "the level of confidence that an entity participating in a network system places on another entity of the same system for performing a given task".

Related to trust, reputation is defined as an objective concept. Jøsang [11] asserts that it is possible to trust someone based on his good reputation or despite his bad reputation; this means that reputation is a factor of trust, but it is not the only one.

As Yan et al. [24] state "trust management plays an important role in IoT for reliable data fusion and mining, qualified services with context-awareness, and enhanced user privacy and information security. It helps people overcome perceptions of uncertainty and risk and engages in user acceptance and consumption of IoT services and applications".

2.2 K-Model

In our previous work [7], we have proposed a framework to consider trust during the whole SDLC of a smart IoT entity. This framework is composed of a K-Model as shown in Fig. 1. Moreover, it includes transversal activities (i.e., Decision-Making, Traceability) belonging to different phases. Furthermore, the context layer is considered for each phase. In fact, we always need to take context into consideration in an IoT environment due to its dynamicity and heterogeneity. However, context can be related to the environment or the rules of the company developing the product.

In the K-Model, we have considered different phases to cover all the SDLC of a smart IoT entity: from cradle to grave. The first phase concerns the need phase, where it is decided the purpose of the new IoT entity and the stakeholders have a key role in it. After this phase, we define the requirements phase, where developers elicit requirements in relation to the previous needs. During the second phase of the K-Model, it is possible to encounter conflicts among requirements. This situation can be due to the needs arisen from different stakeholders. In this paper, we assume that the requirements elicitation process has been completed following the TrUStAPIS methodology [6] and we have to deal only with conflictual requirements. Therefore, decision-making can help in solving this issue. Useful parameters to be taken into consideration are context, traceability and requirements domains, as we will discuss later.

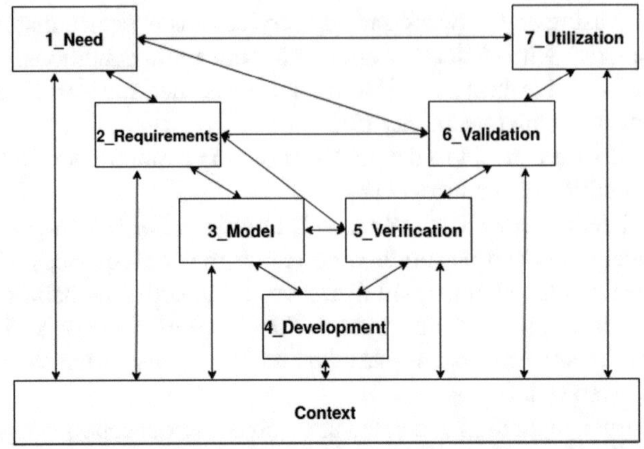

Fig. 1 K-Model: with the *Transversal Activities* it compounds the framework [7]

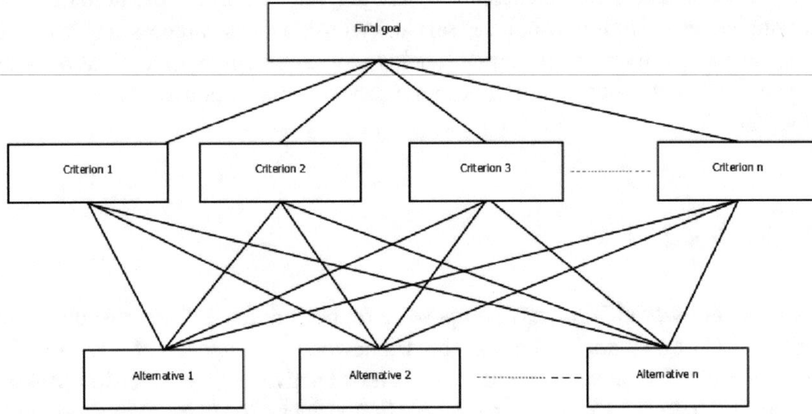

Fig. 2 General AHP model

2.3 Analytic Hierarchy Process (AHP)

The Analytic Hierarchy Process (AHP) has been developed by Saaty [20] and it is a structured technique for organising and analysing complex decisions. It is one of the most widely used methodologies in MCDA [21].

A peculiarity of this methodology is that it is possible to compare among them aspects related to different fields. This is very important because, in accordance with the multidisciplinary aspect of trust and the heterogeneity of IoT, it is useful to be able to compare alternatives from different areas and see which could be the most important with respect to a particular goal.

Basically, AHP permits deciding between various alternatives which one better satisfies a predetermined goal. The alternatives have to be compared with respect to

Table 1 The fundamental scale for pairwise comparisons [20]

Intensity	Meaning
1	A and B are equally important
3	A is relatively more important than B
5	A is more important than B
7	A is much more important than B
9	A is absolutely more important than B

determined criteria. We can see the general AHP model in Fig. 2. The final goal is located at the top layer, at the intermediate layer, there are the criteria and, finally, the alternatives are located at the bottom layer. In the case a higher level of detail is needed, it is possible to divide a criterion into sub-criteria.

The nodes of each layer are pairwise compared with respect to their contribution to the nodes above them. The results of these comparisons are entered into a matrix which is mathematically processed to derive the priorities for all the nodes belonging to that level.

During the pairwise comparisons, we have to decide which element is more important, if the first one (A) or the second one (B). Saaty [20] proposed a fundamental scale to compare the elements. It is shown and explained in Table 1.

When we assign a number to the most important element, according to the fact that the relationship is symmetrical, we have to assign to the other element a symmetrical value. So the more important A is for B, the less important B will be for A. For example, if we have to compare A and B in accordance with a criterion C and we think that A is more important than B, we give A a value of 5 with respect to B, so we have to give B a value of 1/5 with respect to A.

AHP usually starts with comparing the criteria with respect to the goal and the value of the comparison fills the matrix related to them. After that, the DM must compare the sub-criteria with the originating criterion and the values are inserted into the related matrix. Finally, the DM must compare the alternatives with respect to each (sub)criterion; again the values fill the proper matrix.

For every matrix, a Consistency Index (CI) is calculated. It is used to check if the choices made during the comparisons are consistent or not. If CI is lower than 0.10 [20], the matrix can be considered consistent. If not, the DM must reconsider the choices made and solve the consistency issue. The difficulty grows with the number of elements and it is harder to avoid inconsistency. In fact, AHP utilization is discouraged for more than ten elements to be compared [20] because of the risk of inconsistent computing values during the pairwise comparisons. This is an issue that in an environment such as the IoT can limit the effectiveness of this approach. We mitigate this issue with our methodology, as we show in Sect. 3.

2.3.1 AHP in the State of the Art

AHP is applied in individual and group decision-making and it is an effective and flexible tool for structuring and solving complex group decision-making situations, as Altuzarra et al. stated in their work [2].

In Computer Security, Lee [14] has implemented a Fuzzy AHP approach about information security and risk assessment. He mainly considered four criteria: assets, threats, vulnerability and safety measures. Kim [13] has presented an AHP method based on network interfaces and a channel selection algorithm for multichannel MAC protocols in IoT ecosystems that considers several decision-making factors such as expected channel condition, latency and frame reception ratio. The proposed scheme considers an IoT-based healthcare system and can fit in a more complex use case scenario similar to the one that we will propose later in this paper.

AHP has been considered in the state of the art on trust management only in a few works. One of them is the work of Pang et al. [17] where they implement AHP in order to address the fact that IoT nodes have limited computational power. They used the model in order to calculate the trust level by implementing reputation parameters. They do not consider related domains, as privacy or security.

According to requirements elicitation, Taha et al. [21] introduced an AHP-based technique that allows a comparative analysis of cloud security. This is interesting from our perspective because their technique simplifies security requirements specifications. They consider a methodology that helps users to better understand and identify their security needs. However, in our framework [6], we have considered security requirements according to other types of requirements (i.e., availability, privacy) that in their work are missing. Finally, Kassab et al. [12] explore AHP in order to assist the prioritization of quality requirements. This approach can be limited due to the increase of complexity if it is used for all the requirements. However, in our approach, we focus only on the conflictual requirements.

3 Pairwise Ordination Method (POM)

POM is composed of a goal, alternatives and criteria like the AHP, but the operations among these elements are performed in a different way. Anyhow, the goal is to rank the requirements that guarantee the maximum level of trust for the developing IoT entity according to the criteria and the conflicting alternatives. Then, the criteria, as long as for AHP, can be divided into sub-criteria to improve the level of detail. Finally, there are the alternatives that, in our case, are conflicting requirements.

A peculiarity of this methodology is that it is possible to compare aspects related to different fields prioritising them with a normalised value. Thus, we can compare criteria such as time with others like cost and choose between them. This aspect is very important because, in accordance with the multidisciplinary aspect of IoT, it is useful to be able to compare aspects belonging to different areas.

The goal we have to achieve is to choose the requirement that maximizes the trust level of the developed system. To fulfil this trusted goal, we have identified three groups of criteria that are mandatory to be used in our methodology:

1. **Context criteria**. As we stated in [7], the context can be a composition of functionalities or dependent on the environment. Context is always present and it needs to be taken into consideration in an IoT environment. Moreover, these criteria can affect the trust value of the whole system. Furthermore, according to the general aspects related to trust, we have to identify *general criteria* that can affect the trust level of the system.
2. **Traceability criteria**. Traceability is crucial during the development of an IoT entity [7]. Moreover, we cannot solve a conflict among requirements without considering which other elements (i.e., requirements, needs, models or documents) are involved and connected with the conflicting requirements. For this reason, the complexity of the IoT entity under development must be taken into consideration in order to decide how to proceed. In addition, the correct implementation of traceability improves the trust level of the developed IoT entity [6].
3. **TrUStAPIS criteria**. These criteria are related to our requirements elicitation method presented in [6]. Moreover, as we also stated in [7], trust is strongly related to other domains: privacy, identity, security, usability, safety and availability. The domains that must be considered as criteria are the same of the conflicting requirements. There can be a maximum of seven criteria.

We will show how these criteria can be covered in the use case scenario proposed in Sect. 4.

The alternatives that we take into consideration are requirements that cause conflicts among them. Furthermore, we assume that the elicitation process and the identification of these conflicts have already been performed. The criteria must belong to the classification groups we have shown before.

3.1 Procedure

In this methodology, we have to perform comparisons between the elements belonging to contiguous layers. We start comparing goals and criteria. Then, we compare criteria and sub-criteria (if they exist) and finally, we compare the alternatives with criteria or sub-criteria.

These comparisons are necessary to order alternatives and criteria from the less to the more important. At the end of the procedure, we can rank the requirements from the most to the least important. The importance is given to which one of the requirements is the most important in order to improve the level of trust of the system.

During each comparison, we create an ordered branch-tree. For each round, we compare an element with the following not yet ordered element and the DM decides which one is more important in accordance with the criterion or goal. The procedure is iterative and each time an element *wins* the comparison, it has to be compared

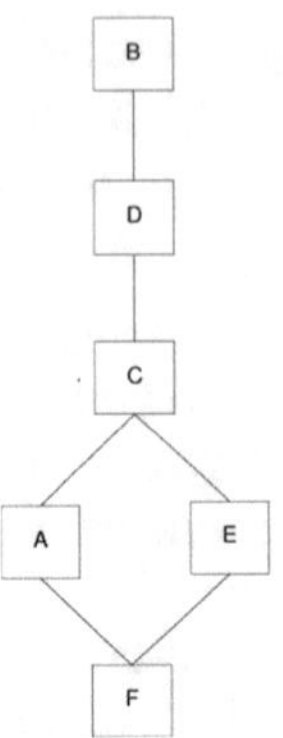

Fig. 3 Example: ordered branch-tree

with the following element. The round ends after each element has been compared at least once. When the first round finishes, we obtain a partially ordered tree. In the following rounds of comparisons, we will compare the not yet ordered elements until we have a completely ordered branch-tree. Anyhow, it is possible that this branch should have some levels populated by more elements.

For example, let us assume that we have six alternatives (A, B, C, D, E and F) and we have to compare them according to criterion X.

In the first round, we have to compare all the elements among them in order to decide which one is the most important. Thus, firstly, we compare A with B in order to decide which alternative is more important. Imagine that the DM decides that B is the most important. It means that B will be then compared with C. Let us assume that B wins all the comparisons until F. Thus, we find out that B is the most important alternative among the others according to criterion X, but we do not know how to rank the other alternatives. So, we have to estimate also the order of the other alternatives and we start the second round of comparisons.

Firstly, we compare A and C and we find out that C is the most important. Secondly, C is compared with D and the DM decides that D is more important than C. Then, we have to compare D with E and D wins. Finally, we compare D with F and D is the most important.

Now, we know that B is the most important element, followed by D. But we have to order also the remaining alternatives.

In the third round, we compare A and C again, but we previously found out that C was more important than A, so it is possible to skip this comparison. Then, we compare C and E deciding that C is the most important. Finally, C is compared with F and C wins again.

Now, our branch-tree is composed of B, C and D ranked in this order. The remaining alternatives to be ordered are A, E and F.

Thus, we start the fourth round by comparing A and E and we find out that they are equally important according to X. Finally, we compare A with F and the DM decides that A is more important.

In the last round of comparisons, we find out that E is more important than F because A and E have the same importance.

So, the algorithm ends and we have an ordered branch-tree. The algorithm representing our case is the following (Fig. 3):

Algorithm 5 Trust model algorithm for home devices

```
procedure POM
    while Each_element_is_not_ordered do
        branch["A", "B", "C", "D", "E", "F"]
        orderedBranch["", "", "", "", "", ""]
        for all i = 1 to branch.size() − 1 do
            if i = 0 then
                Comparethefirsttwoelementsofbrancharray
                if Oneofthetwoelementsisbetter then
                    BestElementWins
                elseTheyareequals
                    BothElementWins
            else
                if Oneofthetwoelementsisbetter then
                    BestElementWins
                else
                    BothElementWin
            Save_Winner(s)_into_orderedBranch
            Remove_Winner(s)_from_branch
        orderedBranch["B", "D", "C", "A, E", "F"]
        branch["", "", "", "", "", ""]
```

Now that the branch-tree is ordered, we need to give a weight to each element in order to be normalized and compared with the other branch-trees related to other criteria. In our approach, the weights are based on the number of elements and on their importance. The maximum weight value is equal to the number of elements, the minimum value is one. In order to assign the weights, we proceed following a bottom-up approach. If the bottom layer has only one element, the element belonging to the layer above will have a value equal to the lower value plus one. In the case a layer has more than one element of the same importance, they will have the same value. Anyway, in this case, the element above them will have a value equal to their value plus the number of equal elements. We use this "jump" to highlight the difference between the upper element with respect to the lower elements.

We can summarize these cases in the following general formula:

$$Element\_weight = Lower\_element\_weight + Number\_of\_lower\_elements$$

We can consider the previous example in order to show how the weights are given.

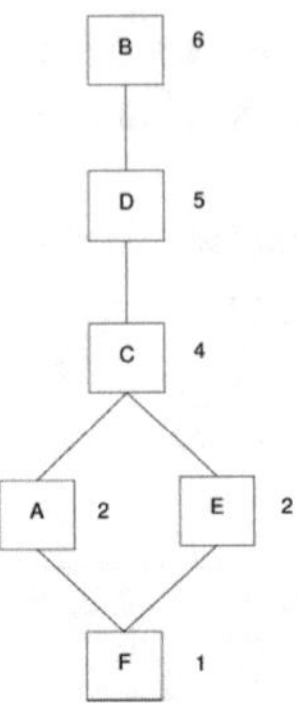

Fig. 4 Example: ordered and weighted branch-tree

We had six elements (A, B, C, D, E and F). So, we have 6 as the maximum value and 1 as the minimum value. So, we give to F weight 1 because it is the lowest. Then, we have two elements of the same importance (A and E) and we give to both of them the weight of 2. Above them, there is C which has a weight of 2 (the value of the element beneath it) plus 2 (the number of elements of the same value in the lower level). So, the weight of C is 4. Then, we have D which is equal to 5. Finally, we have the most important element which has a weight of 6. Our weighted and ordered branch-tree is shown in Fig. 4.

Considering that in a real example there will be other criteria, we have to normalize the alternative values in order to compare them with the results related to the other criteria. This operation will also be performed also for the criteria according to the final goal.

Considering the previous example, to normalize the elements we have to divide each of them for the sum of the weights that is 20. Thus, the normalized values for each alternative will be:

$B = 6/20 => 3/10$
$D = 5/20 => 1/4$
$C = 4/20 => 1/5$
$A = 2/20 => 1/10$
$E = 2/20 => 1/10$
$F = 1/20$

The sum of all the normalized weights is 1. In order to reach the final goal and to choose which alternative is the most important, each alternative will be multiplied for the normalized weight of the sub-criteria and criteria among them and added to the values of the same alternative compared to the other sub-criteria and criteria. The final sum of all these values will be 1 and the higher alternative will be the most important.

We will show this procedure with a complete example in Sect. 4.

Table 2 Maximum comparison related to the number of alternatives. Legends: i = number of alternatives , max = maximum number of comparisons, min = minimum number of comparisons

i	1	2	3	4	5	6	7	n
max	n.a.	1	3	6	10	15	21	$n(n-1)/2$
min	n.a.	1	2	3	4	5	6	$n-1$

3.2 POM Versus AHP

The complexity of POM is the following: the maximum number of comparisons depends on the number of alternatives and the minimum number of comparisons is the number of alternatives minus one, as we show in Table 2. The latter is the best case, in fact, imagine that with each comparison the following element always wins, it means that the previous elements have been already ordered and we need only one round of comparisons to order the branch-tree.

In AHP, we have a fixed number of comparisons that is equal to the maximum number of comparisons of our method:

$$AHP - comparisons = n(n-1)/2$$

A problem that is possible to face using AHP is the increasing possibility to have an inconsistent matrix when the number of elements to be compared grows. With POM, the inconsistency is avoided because the comparison tree is ordered at every step of the algorithm. Another difference between AHP and our methodology is related to the weights. If in AHP, the weights are shown in Table 1, in our approach the weights depend on the number of elements. Thus, in our methodology the weights are fixed. However, even if in AHP the weights can represent better the differences among the elements, this possibility can increase errors due to subjective decisions. This error is mitigated by our approach.

4 Use Case Scenario: Smart Hearth-Monitor

We propose a use case concerning a health service scenario. In this scenario, we assume that during the development of a smart heart monitor an issue has arisen. The issue is due to three needs belonging to different stakeholders creating conflict among them during the requirements elicitation process. The three stakeholders are vendors and customers (patients and doctors) and the elicited requirements follow the TrUStAPIS methodology [6].

The vendors want to know the more data is possible about the customers. For the patients, it is necessary to remain anonymous or at least they need to know that only trusted users can access their data. For the doctors it is necessary to know at least age, gender, diseases and blood type of the user in order to be helpful for the user.

The patients want to be monitored, but they care about privacy and trust in order to submit their personal data. The vendors want to monitor and improve the product in order to sell it. For this reason, they need a huge amount of data from the customers (i.e., age, gender, nationality, address).

We have identified five conflictual requirements and, without solving this issue, it is not possible to continue the development process of the product. The requirements are five: one privacy, one trust, one identity and two availability requirements. They are better explained in Sect. 4.2.

The criteria belong to the ones identified in Sect. 3 and they are defined according to this scenario in Sect. 4.1.

The goal is to decide which requirement to keep that can maximize the level of trust of the IoT entity perceived by the "conflicting" stakeholders.

4.1 Criteria

According to our scenario, we have considered the following important aspects as decision criteria:

- **Context criteria**

  1. **Stakeholders Importance**. When a product is developed, there are different stakeholders that can make decisions and they are very important for the project. In fact, a vendor could stop the project or the customers could not buy the product. For example, if a vendor decides to stop the project, the customers will never have it; on the other hand, if the customers do not buy the product, it will be a failure. In order to be more specific, this criterion needs to be divided into three sub-criteria.

     a. *Vendors*. For the vendors, it is important to have as much information as possible about their customers.
     b. *Doctors*. The doctors need to know all the useful information about the patients to be able to treat their diseases (i.e., blood type, historical diseases).
     c. *Patients*. The patients do not want to share their sensitive information with untrusted users. They can accept at least to share the minimum information as possible with trusted users in order to be properly cured.

  2. **Faster**. It is possible that a requirement could be easier than another in order to be implemented. If a requirement is very hard to be achieved, it can be relaxed in the case of conflicts with other requirements. On the other hand, in the case of a strict deadline, this can be the most important parameter. This criterion can be objective.
  3. **Cheaper**. This criterion is about the cost of the implementation. In the case of conflicting requirements and a limited budget, it can be the most important parameter to be taken into consideration. It is an objective parameter.

- **Traceability criteria**. These criteria are objective. In fact, according to them, the more connections with other elements are developed, the more important the requirement is.

  1. **Connected Requirements**. This information is provided by the traceability database [6]. Thus, if a requirement is released, then the connected requirements can be affected by this operation. The more requirements are connected, the more important the requirement is.

  2. **Connected Needs**. Requirements derive from needs, so this is an important element to be taken into consideration. This knowledge is guaranteed by the documentation activity [7]. Thus, if a requirement is released or must be changed, it is possible to go back to the originating need and change it in accordance with the stakeholders.

- **TrUStAPIS criteria**. These criteria depend on the type of requirements chosen as alternatives. It is important in order to highlight how much a requirement belonging to a particular domain (i.e., privacy) is connected to another domain (i.e., trust). In this use case scenario, the domains are the following: trust, security, privacy and availability.

We need to compare the main criteria according to the goal and the sub-criteria according to the originating criterion. We show the full process in Sect. 5.

4.2 Alternatives

The alternatives are the following:

1. **Privacy Requirement**. PRIV01—The customer shall remain anonymous.
2. **Availability Requirement**. AVBT01—The doctor shall access patients' information related to his/her pathology.
3. **Availability Requirement**. AVBT02—The vendor shall access to all customer information.
4. **Identity Requirement**. IDNT01—The customer shall provide his/her personal data to the system.
5. **Trust Requirement**. TRST01—The customer data shall be accessed only by trusted users.

These requirements create conflict among them because they belong to different stakeholders with incompatible needs. Our methodology helps to choose the most important requirement in order to assure the highest possible trust value on the IoT entity. In some cases, changing the requirements means that the originating need must be changed as well.

As we explained earlier, the alternatives must fulfil the goal. In this scenario, it is important to rank the requirements in order to release or change the less important requirements and keep the others avoiding conflicts.

Now that we have presented criteria and alternatives, we can build the POM model according to the use case scenario proposed. It is shown in Fig. 5 where, for the sake of simplicity, we represent the connection among the alternatives and the criteria with a single point of contact. The dotted line represents a many-to-many connection among them.

5 Results and Discussion

We start the process by comparing the criteria according to the final goal. Secondly, we will compare the sub-criteria to their main criterion. Finally, we will compare the alternatives to their main (sub)criterion. Each section is named as the element considered for the comparisons. For the sake of simplicity and space limitation, we will not represent all the rounds of comparisons in all of the following sections, but we have used the methodology we have shown earlier.

5.1 Goal

We can state that the goal is reached considering the following elements:

$$Goal = \{Context, Traceability, TrUStAPIS\}$$

We need to create a ranked order among criteria to show which of them is the most important according to the goal. Thus, following our methodology, we start comparing the Context (Cx) with the Traceability (Ty). We decide that Cx is the most important, so then we have to compare Cx with TrUStAPIS (Ts) and we decide that Cx is more important than Ts. For both the decisions, we consider the context as crucial, because as presented in [7], it is an element always present and strictly connected to a particular parameter. After the first round, we know that Cx is the most important element. We need another round in order to decide between Ty and Ts. We decide that Ty is the most important because the more requirements or needs are connected, the most important is the requirement.

Thus, we will have the following normalized values:

Cx = (1/2)
Ty = (1/3)
Ts = (1/6).

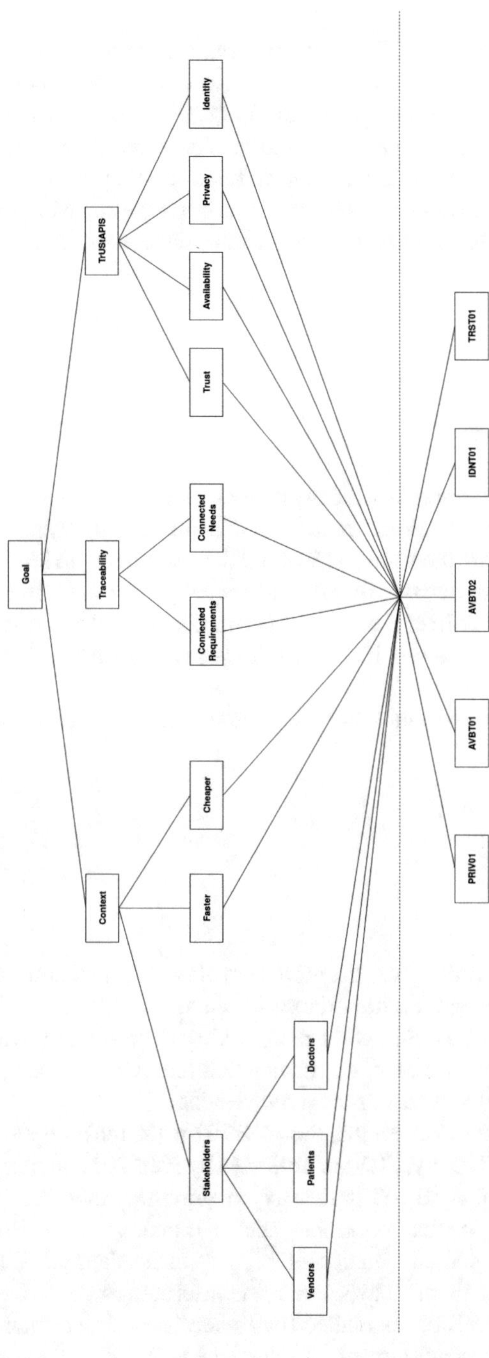

Fig. 5 POM model related to our use case scenario

5.2 Context

The context is composed of three sub-criteria: the stakeholders (Sk) criterion and others strictly dependent on the implementation of the requirements: faster (Fs) and cheaper (Ch). We need to compare them in order to decide which one is the most important. Comparing the stakeholders to the faster and cheaper criteria, we decide that the stakeholders are the most important because they are the ones who provide the needs. Then, we give faster and cheaper criteria the same importance.

Thus, the values related to the sub-criteria of the context are:
Sk = (3/5)
Fs = (1/5)
Ch = (1/5).

5.2.1 Stakeholders

Stakeholders are very important for any project. They are the actors having an interest in the system. We have identified three main stakeholders strictly related to the conflict requirements. They are the vendors (Vn) and the customers (divided into doctors (Dr) and patients (Pt)). We decided that they are equally important. In fact, it is true that if the vendors do not deliver the product, it will not be used by the customers. On the other hand, it is also true that if the customers will not trust and buy the product, it will be a failure.

For these reasons, the values are the same for all the stakeholders:
Vn = (1/3)
Pt = (1/3)
Dr = (1/3).

Vendors

We can state that the vendors are the producers of the IoT device and the requirements are ordered considering their importance for them.

The best way to make this order is by asking them directly which requirement they prefer, but in this use case, we assume that the developer can perform this task by analyzing the collected needs and requirements.

For the vendors, we find out that the AVBT02 is the most important requirement. Secondly, IDNT01. Thirdly, TRST01 followed by AVBT01 and finally PRIV01. This order is due because AVBT02 is the only requirement that directly considers the vendors and it is the one that most represents their interests in the IoT entity. Then, in order to have the needed information, IDNT01 satisfies their need. Thirdly, TRST01 can be important for them to avoid that this information could be manipulated by malicious entities. AVBT01 is ranked fourth because at least some information can be available with this requirement. The last one is PRIV01, because if the patients and doctors remain anonymous, the vendors will not have valuable information for market purposes.

The normalized values are the following:
AVBT02 = (1/3)
IDNT01 = (4/15)
TRST01 = (1/5)
AVBT01 = (2/15)
PRIV01 = (1/15).

Patients

For the patients, the most important requirement is PRIV01, because it guarantees them to be anonymous. Secondly, TRST01 can be a good compromise because at least they know that only a trusted user can manipulate their information. The third one is AVBT01, because in case anonymity is not possible, they accept that the doctors can access their health information in order to be properly cured. Then, IDNT01 can be accepted if the previous requirements are kept. Finally, AVBT02 is the last choice because they do not want the vendors could access all their information.

Thus, the rounds of comparisons produce these normalized values:
PRIV01 = (1/3)
TRST01 = (4/15)
AVBT01 = (1/5)
IDNT01 = (2/15)
AVBT02 = (1/15).

Doctors

The doctors require patients' health information in order to effectively cure them. For this reason, the most important requirement for them is AVBT01. Then, TRST01 is important because also their data must be inserted into the system and they want only trusted users to access them. IDNT01 is the third one because it is important in order to have patients' data. The fourth one is PRIV01 and the fifth is AVBT02. Both of them are not important for the doctors, but they at least prefer PRIV01 because they can remain anonymous in the system and patients' diseases will be revealed, respecting their privacy.

According to the doctors, the final values are:
AVBT01 = (1/3)
TRST01 = (4/15)
IDNT01 = (1/5)
PRIV01 = (2/15)
AVBT02 = (1/15).

5.2.2 Faster

As for the Stakeholders criteria, this criterion and the following one are strictly dependent on the context. The more a requirement is simple to implement, the faster it is. For this reason, after the first round of comparisons, we have identified two equally important requirements: AVBT02 and IDNT01. In fact, these two requirements do not need any filter. They are simple operations of reading and writing data in the database. Then, AVBT01 is more difficult to be implemented because it requires a specific filter. The fourth one is TRST01 because it needs a trust decision model or an authentication process in order to be implemented. Finally, the slower requirement to be implemented is PRIV01 because it requires to anonymize the data.

According to the faster sub-criterion, the final values are:
AVBT02 = (2/7)
IDNT01 = (2/7)
AVBT01 = (3/14)
TRST01 = (1/7)
PRIV01 = (1/14).

5.2.3 Cheaper

This criterion is very important in the case the budget is limited or the stakeholders want to maximise the income. However, in the early phases of the SDLC, it is possible that even if developers and stakeholders avoid implementing an expensive requirement, there is the possibility that some issues arise in the following phases of the SDLC. In this case, the amount of money spent to solve the issues will be higher than the money that would be spent for the original expensive requirement [8].

Anyhow, in our case, after the first round of comparisons, IDNT01 is considered the cheapest requirement to be developed. Then, AVBT02 and TRST01 are equally considered. Finally, there are AVBT01 followed by PRIV01. The last one is both the slower and the more expensive because implementing anonymity is the hardest task considering the other requirements. IDNT01 is considered the cheapest because it is the basic requirement to be implemented.

The normalized values are:
IDNT01 = (5/14)
TRST01 = (3/14)
AVBT02 = (3/14)
AVBT01 = (1/7)
PRIV01 = (1/14).

5.3 *Traceability*

Traceability is a transversal activity of the K-Model [7] and it is very important in order to connect requirements among them. In our scenario, there are two sub-criteria. They are related to the connected requirements (Con_R) and the connected needs (Con_N). We decide to give more importance to Con_N because the needs are the real motivation behind a product, so if a requirement is strictly connected to a need, it should be more important than another one only connected to a requirement.

The values related to them are:
Con_N = (2/3)
Con_R = (1/3).

5.3.1 Connected Requirements

This criterion can be completely objective. The operation needed is to count how many requirements are connected to the requirement under consideration. In this use case, we have to subjectively consider the possible relationships among the requirements.

Thus, because we consider trust as central and more connected to the other domains, we consider it as the most connected requirement. Then, we consider that AVBT01 should be the second one because it also refers to the patients and to a particular type of data. Then, we consider AVBT02 and IDNT01 equally important. Finally, there is PRIV01. We assume that even if it is the harder requirement to be implemented, it does not need any other requirements to be connected with in order to be elicited and developed.

Thus, the final normalized values about Con_R are:
TRST01 = (5/14)
AVBT01 = (2/7)
AVBT02 = (1/7)
IDNT01 = (1/7)
PRIV01 = (1/14).

5.3.2 Connected Needs

In this case, it is important to consider how many needs are connected to a single requirement. It is possible that more needs originate from a single requirement or even that a requirement is not connected to any need but only to other requirements (especially if it is a sub-requirement [6]).

In the case of a real use case scenario, it is possible to count all the connected needs and apply our methodology in an objective way. However, in this case, we proceed as for the previous element.

After the first round of comparisons, we decided that IDNT01 is the requirement connected to more needs because it is connected to much user information. For this reason, numerous needs shall be connected to it. The second requirement is PRIV01. In fact, we assume that not only the customers require it but also data protection regulations needs are connected to the privacy requirement (i.e., GDPR [23]). The third requirement is TRST01. Then, AVBT02 has more connected needs than AVBT01, because we imagine that the vendors satisfy more needs than the doctors obtaining all the customer information.

The final values are:
IDNT01 = (1/3)
PRIV01 = (4/15)
TRST01 = (1/5)
AVBT02 = (2/15)
AVBT01 = (1/15).

5.4 TrUStAPIS

In order to decide which requirement to keep, it is important to study the relationships among the requirements domains and the conflicting requirements. This criterion gives a holistic view of how a requirement of a particular domain can be related to others.

In this case, we compared the trust (Tt) domain firstly with the availability (Av) domain and we gave more importance to trust. Secondly, to the privacy (Pr) domain and again we decide that trust is more important (according to our particular goal). Finally, we compare it to the identity (Id) domain giving more importance to trust.

During the second round, we compare availability to privacy, deciding that the latter is more important. Then, we compare the privacy requirement to the identity requirement and we decide that they are equals. For this reason, there is no need to proceed with a third round of comparison. In fact, privacy is more important than availability and consequently because identity is equal to privacy, identity is also more important than availability.

Thus, the values related to the TrUStAPIS sub-criteria are:
Tt = (4/9)
Id = (2/9)
Pr = (2/9)
Av = (1/9).

5.4.1 Trust

This criterion is related to trust and we want to prioritize all the requirements according to this domain. Thus, the most important requirement is TRST01. Then, we have IDNT01, because, in order to perform this operation, the customers must trust the system. Otherwise they will not use it. Then, AVBT02 is another requirement that implies a high level of trust in the system. Because only if the customers can trust the vendors it should be possible to implement it. Then, there is AVBT01 which is related to the trust between patients and doctors. Finally, there is PRIV01 that must be implemented if there is a low level of trust between customers and vendors.

The final normalized values are:
TRST01 = (1/3)
IDNT01 = (4/15)
AVBT02 = (1/5)
AVBT01 = (2/15)
PRIV01 = (1/15).

5.4.2 Availability

After the two availability requirements (equally important), the third one is IDNT01, because only if the customers provide their information, they will be available. Then, TRST01 is important for availability, because it guarantees that the data are available for trusted users. Finally, PRIV01 is the least important requirement according to availability, because if the data are anonymous, they could not be easily available.

For availability, the normalized values are:
AVBT01 = (2/7)
AVBT02 = (2/7)
IDNT01 = (3/14)
TRST01 = (1/7)
PRIV01 = (1/14).

5.4.3 Privacy

According to privacy, the most important requirement is PRIV01. Then, the second one is AVBT01, because it guarantees that only the doctors can access sensitive information. TRST01 is the third parameter because it guarantees that only trusted users can access the data. Then, there are both AVBT02 and IDNT01.

The final values are:
PRIV01 = (5/14)
AVBT01 = (2/7)
TRST01 = (3/14)
IDNT01 = (1/14)
AVBT02 = (1/14).

5.4.4 Identity

After the identity requirement, the second one is AVBT02 because it requires that the identity data should be available for the vendors. The third requirement is AVBT01 because it provides at least some data for the doctors and they need to know the identity of the patients in addition to their diseases. Then, there is TRST01 and finally PRIV01. The latter does not provide any identity information.

The final normalized value about identity are:
IDNT01 = (1/3)
AVBT02 = (4/15)
AVBT01 = (1/5)
TRST01 = (2/15)
PRIV01 = (1/15).

5.5 *Final Priority*

After the calculation of the normalized local weights, we have to compute the final priority related to each of the alternatives according to the goal. In order to perform this activity, we must sum every value related to the single alternatives by multiplying it for each value of the sub-criteria and criteria above them.

Because the values are normalized, the sum of the final results will be 1.

In Fig. 6, there are the values derived from Sects. 5.1 to 5.4.4. They are the same calculated earlier.

In this figure, to avoid having many lines and boxes for the alternatives, we have summarized them in a single box. P1 is the privacy requirement, A1 and A2 are the availability requirements, I1 is the identity requirement and T1 is the trust requirement.

Thus, in order to calculate the single priority related to the conflicting requirements, starting from P1, we will have.

$$\mathbf{P1} = (1/15) * (1/3) * (3/5) * (1/2) + (1/3) * (1/3) * (3/5) * (1/2) + (2/15) * (1/3) * (3/5) * (1/2) + (1/14) * (1/5) * (1/2) + (1/14) * (1/5) * (1/2) + (1/14) * (1/3) * (1/3) + (4/15) * (2/3) * (1/3) + (1/15) * (4/9) * (1/6) + (1/14) * (1/9) * (1/6) + (5/14) * (2/9) * (1/6) + (1/15) * (2/9) * (1/6) = \mathbf{0.157}$$

For the other requirements, we have:

$$\mathbf{A1} = (2/15) * (1/3) * (3/5) * (1/2) + (1/5) * (1/3) * (3/5) * (1/2) + (1/3) * (1/3) * (3/5) * (1/2) + (3/14) * (1/5) * (1/2) + (1/7) * (1/5) * (1/2) + (2/7) * (1/3) * (1/3) + (1/15) * (2/3) * (1/3) + (2/15) * (4/9) * (1/6) + (2/7) * (1/9) * (1/6) + (2/7) * (2/9) * (1/6) + (1/5) * (2/9) * (1/6) = \mathbf{0.182}$$

$$\mathbf{A2} = (1/3) * (1/3) * (3/5) * (1/2) + (1/15) * (1/3) * (3/5) * (1/2) + (1/15) * (1/3) * (3/5) * (1/2) + (2/7) * (1/5) * (1/2) + (3/14) * (1/5) * (1/2) + (1/7) *$$

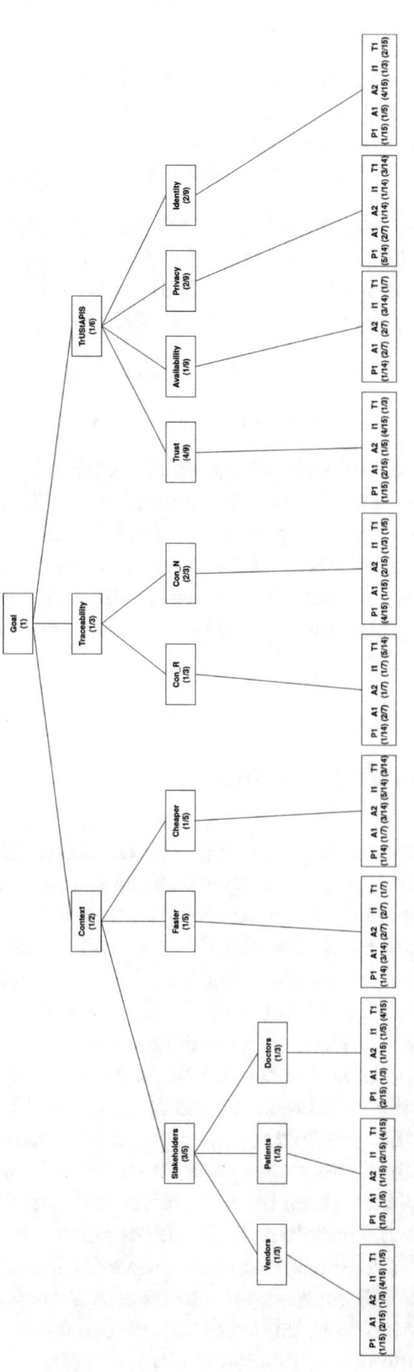

Fig. 6 POM model related to our use case scenario

$$(1/3) * (1/3) + (2/15) * (2/3) * (1/3) + (1/5) * (4/9) * (1/6) + (2/7) * (1/9) * (1/6) + (1/14) * (2/9) * (1/6) + (4/15) * (2/9) * (1/6) = \mathbf{0.175}$$

$$\mathbf{I1} = (4/15) * (1/3) * (3/5) * (1/2) + (2/15) * (1/3) * (3/5) * (1/2) + (1/5) * (1/3) * (3/5) * (1/2) + (2/7) * (1/5) * (1/2) + (5/14) * (1/5) * (1/2) + (1/7) * (1/3) * (1/3) + (1/3) * (2/3) * (1/3) + (4/15) * (4/9) * (1/6) + (3/14) * (1/9) * (1/6) + (1/14) * (2/9) * (1/6) + (1/3) * (2/9) * (1/6) = \mathbf{0.253}$$

$$\mathbf{T1} = (3/15) * (1/3) * (3/5) * (1/2) + (4/15) * (1/3) * (3/5) * (1/2) + (4/15) * (1/3) * (3/5) * (1/2) + (1/7) * (1/5) * (1/2) + (3/14) * (1/5) * (1/2) + (5/14) * (1/3) * (1/3) + (1/5) * (2/3) * (1/3) + (1/3) * (4/9) * (1/6) + (1/7) * (1/9) * (1/6) + (3/14) * (2/9) * (1/6) + (2/15) * (2/9) * (1/6) = \mathbf{0.233}$$

The results are rounded to three digits after zero. The total is:

P1 + A1 + A2 + I1 + T1 = 1

In conclusion, the most important requirement is IDNT01 followed by TRST01. On the other hand, the least important requirement is PRIV01. For this reason, PRIV01 will be the requirement released or modified in order to solve the conflict among requirements. If this operation does not solve the issue, it is possible to repeat the calculations (in the case the privacy requirement has been modified) or to select the second least requirement (AVBT02) and delete or modify it. The process is iterative until the conflicts are solved.

6 Conclusion and Future Work

Our use case scenario shows the potential of a process AHP-like in order to assist in the decision-makers. We think that our approach can be useful in an IoT environment because of the heterogeneity and dynamicity of this field. In addition, it can be very helpful in later phases of the SDLC, as the utilization [7]. Because of the low computational power of the smart objects, a low computation effort guaranteed by our methodology can be helpful. Even if the subjective analysis is known to have issues, especially if performed by different users, we mitigate this situation by considering only specialized actors performing the comparisons belonging to their field (i.e., a specific stakeholder or the developers). In addition, if there are objective parameters, it is possible to directly consider them in order to perform the comparisons. Another important aspect to be taken into consideration is the increasing of the complexity when the alternatives and criteria grow, as also stated by Metcalfe [16]. This methodology limits the computational effort required and avoids the problems that AHP has when the complexity increases [20].

As future work, we will compare the results of our methodology against AHP in the same scenario. Moreover, this work will be extended to a real use case scenario. Through this approach, comparative studies might be conducted to analyse similarities or differences among trust and reputation models.

Acknowledgements This work has received funding from the NeCS project by the European Union's Horizon 2020 research and innovation programme under the Marie Sklodowska-Curie grant agreement No. 675320, the CyberSec4Europe project under SU-ICT-03 programme grant agreement 830929, and the EU project H2020-MSCA-RISE-2017 under grant agreement No. 777996 (Sealed-GRID).
This work reflects only the authors' view and the Research Executive Agency is not responsible for any use that may be made of the information it contains.

References

1. Agudo I, Fernandez-Gago C, Lopez J (2008) A model for trust metrics analysis. In: International conference on trust, privacy and security in digital business. Springer, pp 28–37
2. Altuzarra A, Moreno-Jiménez JM, Salvador M (2007) A Bayesian priorization procedure for AHP-group decision making. Eur J Oper Res 182(1):367–382
3. Cabała P (2010) Using the analytic hierarchy process in evaluating decision alternatives. Oper Res Decis 20(1):5–23
4. Erickson J (2009) Trust metrics. In: International symposium on collaborative technologies and systems, 2009, CTS'09. IEEE, pp 93–97
5. Fernandez-Gago C, Moyano F, Lopez J (2017) Modelling trust dynamics in the internet of things. Inf Sci 396:72–82
6. Ferraris D, Fernandez-Gago C (2019) TrUStAPIS: a trust requirements elicitation method for IoT. Int J Inf Secur 1–17
7. Ferraris D, Fernandez-Gago C, Lopez J (2018) A trust by design framework for the internet of things. In: NTMS'2018—security track (NTMS 2018 security track). Paris, France
8. Friedenthal S, Moore A, Steiner R (2014) A practical guide to SysML: the systems modeling language. Morgan Kaufmann
9. Gambetta D et al (2000) Can we trust trust. Trust: making and breaking cooperative relations. 13:213–237
10. Hoffman LJ, Lawson-Jenkins K, Blum J (2006) Trust beyond security: an expanded trust model. Commun ACM 49(7):94–101
11. Jøsang A, Ismail R, Boyd C (2007) A survey of trust and reputation systems for online service provision. Decis Support Syst 43(2):618–644
12. Kassab M, Kilicay-Ergin N (2015) Applying analytical hierarchy process to system quality requirements prioritization. Innov Syst Softw Eng 11(4):303–312
13. Kim B, Kim S (2017) An AHP-based interface and channel selection for multi-channel mac protocol in IoT ecosystem. Wirel Pers Commun 93(1):97–118
14. Lee MC (2014) Information security risk analysis methods and research trends: AHP and fuzzy comprehensive method. Int J Comput Sci Inf Technol 6(1):29
15. McKnight DH, Chervany NL (1996) The meanings of trust
16. Metcalfe B (1995) Metcalfe's law: a network becomes more valuable as it reaches more users. Infoworld 17(40):53
17. Pang XY, Wang C (2014) The study of trust evaluation model based on improved AHP and cloud model in IoT. In: Advanced materials research, vol 918. Trans Tech Publ, pp 258–263
18. Pavlidis M (2011) Designing for trust. In: CAiSE (doctoral consortium), pp 3–14
19. Roman R, Najera P, Lopez J (2011) Securing the internet of things. Computer 44(9):51–58
20. Saaty TL (1980) Analytic hierarchy process. Wiley Online Library
21. Taha A, Trapero R, Luna J, Suri N (2014) AHP-based quantitative approach for assessing and comparing cloud security. In: 2014 IEEE 13th international conference on trust, security and privacy in computing and communications (TrustCom). IEEE, pp 284–291

22. Vaidya OS, Kumar S (2006) Analytic hierarchy process: an overview of applications. Eur J Oper Res 169(1):1–29
23. Voigt P, Von dem Bussche A (2017) The EU general data protection regulation (GDPR). A practical guide, 1st edn. Springer International Publishing, Cham
24. Yan Z, Zhang P, Vasilakos AV (2014) A survey on trust management for internet of things. J Netw Comput Appl 42:120–134

Trust Negotiation and Its Applications

Martin Kolar, Carmen Fernandez-Gago, and Javier Lopez

Abstract Trust negotiation is an approach for establishing trust between various entities that is especially useful in online environments. Entities may need to or want to establish a trust relationship in order to reach their goal, on which they can collaborate together. Its principle is to mutually and alternately exchange credentials between the participating entities so that their trust placed onto each other can gradually increase. Trust negotiation must also handle other issues that are important for its functionality. The first and the most important one is security. Entities must be securely authenticated in order to know who they communicate with. Also, they must be authorised for accessing confidential information about the other one, which is handled by specific rules defined in policies. Credentials themselves must be protected too in order to maintain their integrity and prevent malicious modifications and leakage. Privacy is another related issue. It is not always indispensable to protect privacy in order to establish trust, however, entities may be willing to do so. Trust negotiation may be required for many different scenarios and its processing can be carried out automatically without an active intervention from its participants. In this chapter, we analyse trust negotiation in terms of its functionality and possible implementations. We propose a way of how to design trust negotiation effectively for several scenarios since each of them may have different requirements for trust negotiation, such as the needs for security and privacy, the use of various negotiating strategies or the use of trusted authorities. We propose the way of how trust negotiation can be adapted for the specific needs required by its participants.

M. Kolar (✉) · J. Lopez
Department of Computer Science, University of Malaga, Malaga, Spain
e-mail: kolar@lcc.uma.es

J. Lopez
e-mail: jlm@lcc.uma.es

C. Fernandez-Gago
Department of Applied Mathematics, Malaga, Spain
e-mail: mcgago@lcc.uma.es

T. Dimitrakos et al. (eds.), *Collaborative Approaches for Cyber Security in Cyber-Physical Systems*, Advanced Sciences and Technologies for Security Applications,
https://doi.org/10.1007/978-3-031-16088-2_8

1 Introduction

The modern world is highly demanding for information exchange of all types. People, companies, institutions and all the other entities need to share their knowledge, goods and others in order to achieve their goals. It is important that all entities could enjoy fair conditions and were justly rewarded for their provided resources. This requirement is difficult to ensure, however, trust relationships can greatly help to make the entities confident about the others. Traditionally, trust is acquired based on experience during a relatively long period of time and as a result, it can turn positive or negative, so called distrust. For the needs of our modern society, this approach is not always applicable and more advanced techniques are required. Trust negotiation can establish trust relationships between entities on demand by exchanging their credentials. It has advantages over the traditional way as the entities can be total strangers, can negotiate together over a network and the process can be fast. Additionally, trust negotiation can be automated and carried out on behalf of the entities by their security agents [23]. It is especially suitable for online environments, where various applications can build trust automatically and authorise to Web services, etc. Also, the number of participating entities on the Internet is enormous and trust negotiation may help to find and filter a suitable entity for reaching the given goal.

In this chapter, we analyse the concept of trust negotiation and demonstrate several scenarios with different requirements. Then, we propose a trust negotiation model that aims to be as general as possible, covering all the presented scenarios, whereas only the required features can be implemented.

The rest of this chapter is organised as follows. Section 2 presents previous work on trust, trust negotiation, trust models and policies. Section 3 presents the trust negotiation concept and identifies its important criteria. Also, it delves into the access control approaches and definition of policies required for trust negotiation. In Sect. 4, several trust negotiation scenarios are modelled and analysed. Based on the identified criteria, we introduce a generally usable trust negotiation model in Sect. 5. Finally, Sect. 6 concludes the chapter.

2 Related Work

Trust is an important issue for entities when they relate or collaborate together. Computer Science also analyses and utilises the concept of trust as it is meaningful for many areas. The topic of trust is surprisingly wide and there exist many diverse trust definitions, because it is difficult to specify a single general one dealing with all its aspects. It is also easily confused with the other similar concepts, such as credibility, reliability and confidence [21]. Gambetta [8] presents a simple definition, when he claims that trust is a subjective probability, by which an individual A expects another individual B to perform a given action, on which its welfare depends. It is supposed that the entity placing trust (*trustor*) on the other one (*trustee*) is dependent on it and the trustee is reliable. Falcone and Castelfranchi [7] claim that even when an entity highly trusts another one, it might not be generally enough for the trustor to become dependent on the trustee. Jøsang [12] recognises two types of trust definitions called *reliability trust* and *decision trust*. The former can be interpreted as the reliability of something or somebody. This includes the trustor's dependency on the trustee and the trustee's reliability as seen by the trustor. The latter is inspired by the work of McKnight and Chervany [16] and can be understood as a decision of the trustor, to which extent is willing to become dependent on the trustee in a given scenario. It is desirable that the trustor feels secure and acceptable about the trustee, however it should be aware of potential risks. Grandison and Sloman [9] present a more specific definition of trust that is already context-aware: "Trust is the firm belief in the competence of an entity to act dependably, securely, and reliably within a specified context." The context represents a very important part of trust, because trustees are trusted for their attributes that may be valid only within a certain context.

These definitions can be considered as traditional ones that are applied for general scenarios. A more specific concept of trust is used for online systems. It handles trust relationships between entities on the Internet, such as customers and e-commerce web sites. Corritore et al. [6] claim that trust between an individual and a specific website is: "an attitude of confident expectation in an online situation of risk that one's vulnerabilities will not be exploited". The online trust is more vulnerable than the traditional one since it takes place in a complex and anonymous area. Entities do not know each other, they can fake their identity and attributes and their behaviour is unpredictable. For this reason, Tsiakis and Sthephanides [20] suggest to create a trusted and secure environment, in which the entities can be more confident. They present important requirements for establishing it in order to carry out electronic payments, such as: the involved entities must be uniquely identifiable, their trust placed onto the others is unquestionable and only a minimum number of them place trust by default. Trust is made in a process called trust establishment. It may exist as one-direction relationship, where an entity places trust onto another one, or two-direction relationship, where trust is placed reciprocally between them. Winsborough et al. [23] identify two common approaches that presume a previous experience with entities: the *identity-based* approach authenticates an entity based on its known identity and the *capability-based* approach makes authentications based on the entity's capabilities as demanded by a requester. Due to the complex and anonymous character of open systems, these approaches are not suitable. In online environments, trust is established by a mutual exchange of credentials between entities.

The trust interactions can be abstracted and implemented in a trust model. It allows us to control and observe, how trust relationships are created and evolving. The trust models are usually utilised to establish a trusted environment for handling trust relations and access control. Lara [17] classifies them into two categories: *decision models* and *evaluation models*. The former make use of policies and credentials in order to control access and unify authentication and authorisation into one step, which simplifies trust decisions. *Policy models* and *negotiation models* belong to this category. The latter take into consideration diverse factors that participate on the trust relationships. These factors are evaluated and trust is computed based on them. *Propagation models* and *reputation models* belong here. Apart from these, a trust management system (*TMS*) also handles trust dynamics and processes symbolic representations of social trust. It is useful for making automated access control decisions in various information systems. In this chapter, we focus on the trust negotiation models. The first one was TrustBuilder and its extended version TrustBuilder2 [18, 24]. They represent a flexible and fully reconfigurable framework that facilitates trust negotiation for entities. The process of establishing trust can be delegated to security agents that handle all the important aspects on their own. Another one, PROTUNE [5], is rule-based and designed for the client-server architecture. A security agent plays one of the two roles: as a client makes requests to a server and as a server provides resources and evidences to the client. Trust negotiation builds trust between two entities by exchanging their credentials. The revealed information increases credibility of its owner. Trust negotiation overcomes the traditional approaches for building trust as it is versatile, suitable for online environments and does not need previous interactions nor experience between its participants. Cassandra [1] is a trust management system designed for large networks. Each negotiator owns one Cassandra instance that acts as a service and creates a protective layer around its resources. Credentials may be exchanged only through its controlled interface. Cassandra facilitates access control and authorisations for security-critical actions. Another one is PolicyMaker [3, 4]. Its main purpose is the processing of queries of an entity that check for access rights to perform a trusted action. PolicyMaker provides a recommendation based on the entity's credentials, policies and action descriptions. KeyNote [2] represents a simple and flexible TMS suitable for Internet-based applications. It provides a unified language for credentials and policies that form assertions. In fact, they are small programs describing trusted actions that can be sent over the network and delegate control access decisions.

Many trust models, among which the trust negotiation ones also belong, use policies mainly to control access to resources. A policy consists of a group of rules that are combined by logical operators forming expressions. Generally, policies aim to maintain security and are classified based on their functions into categories, such as the privacy, purpose and access control ones. Policies define, which entities are entitled to access credentials, perform trusted actions and under which conditions. It is a good practice to grant additional privileges based on new acquired information and not to revoke them [19]. Policies are defined in a policy language that can use various approaches, such as the attribute-based access control (*ABAC*) [22] or the role-based access control (*RBAC*) [1, 10]. They make use of the general attribute exchange protocols for their implementation: the Extensible Application Markup Language (*XAML*) [15] or the Security Assertion Markup Language (*SAML*) [11]. The most suitable policy language that fits the best our identified criteria for trust negotiation [13] is PlexC [14].

3 Trust Negotiation

In this section, we analyse the trust negotiation principle and describe its basic functionality. We summarise the identified criteria for trust negotiation presented in our work [13]. Then, we analyse the concept of negotiation strategies and overview some of them. We also summarise the most common access control approaches and address a definition of access control policies. We give an overview about the PlexC policy language that seems to be the most suitable for trust negotiation.

3.1 Concept

Trust negotiation is a suitable approach for establishing trust between two entities that want to establish a trust relationship in order to achieve a common goal. It is based on the concept of exchanging credentials between its participants, also called negotiators for the purposes of trust negotiation. Considering the negotiators as total strangers, we can assume they have zero trust to each other and they have to build their trust relationship from scratch. One of the main advantages of trust negotiation is its ability to neither assume previous knowledge nor experience of the participants. In order to perform trust negotiation, entities must possess credentials that are interesting for the other party for the purposes of building trust and they must be plausible. In case of doubts, a third party can be helpful, such as a trusted authority that can confirm their validity or genuineness. Entities must consider their requirements, offers and goals and based on them define their policies. Trust negotiation is policy-driven, so the policies should reflect the actual needs of entities, such as their security, privacy and intentions.
The basic principle of how trust negotiation works is depicted in Fig. 1. Two entities marked as (1) and (2) represent the negotiators that want to establish a trust relationship. Both possess credentials and define their policies. Let us assume that the entities are strangers and therefore their placed trust onto each other is zero. The negotiator (1) begins trust negotiation by requesting one credential from the other

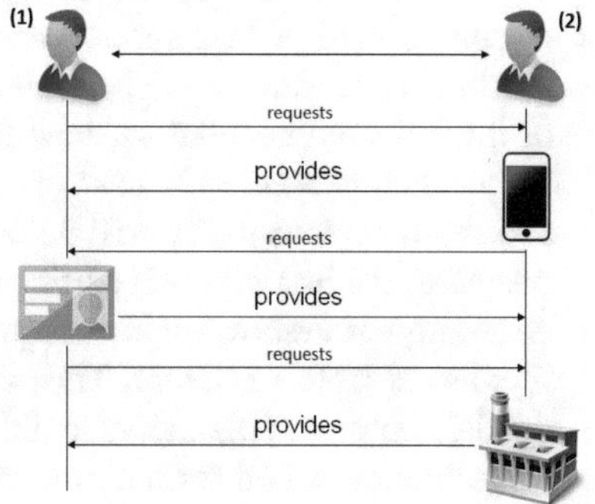

Fig. 1 Trust negotiation basic concept

negotiator (2). The negotiator (2) has to check his policies if he can disclose it and if so, he will provide the credential to the negotiator (1). As the negotiator (1) receives the credential, his placed trust onto the negotiator (2) is increasing and as a result, he might be more willing to disclose his own credentials. Then, the negotiator (2) requests a credential from the negotiator (1) and if the policies are accomplished, the negotiator (1) provides the credential to the negotiator (2). This way, the exchange process of credentials continues and trust is being built by the negotiators mutually. Trust negotiation finishes successfully, if the required trust level has been reached by one negotiator or both of them, depending on the scenario. In case, that not enough trust has been built and one or both of the negotiators are not willing to provide more credentials, trust negotiation terminates with failure.

3.2 *Criteria*

As the trust negotiation concept denotes, building trust is a matter of exchanging credentials and confidential information between two negotiators representing entities. Kolar et al. [13] identify general requirements for trust negotiation in their work. They ensure a secure and reliable exchange process and protect privacy of the negotiators. The following requirements were identified:

- **Privacy of resources**. The entities disclose confidential information that are addressed only to their recipients. An entity must be sure that it will not leak to an unauthorised party nor will be obtained by a swindle. The exchange channel must be secure, so that the exchanged data will remain consistent and unmodified by an attacker. Also, entities must consider the disclosure of their policies as they may lead to discover a strategy for obtaining their credentials unjustly.
- **Access control to resources**. The entities need to define and control access to their credentials, so that only the authorised parties can obtain them. To do so, policies handling access rights must be specified. They contain conditions and actions or delegations, where a condition must be satisfied in order to perform the associated action or delegation. This way, an entity can define a decision tree of disclosures for its credentials. This requirement is related to the previous one as the privacy protection is handled by access control too.
- **Usage control of resources**. The entities are naturally exposed to a risk when sharing their credentials. Therefore it may be suitable to monitor the future actions of the involved negotiators, how they use the obtained credentials and whether follow the given rules. In case of their abuse, trust negotiation will be terminated and the malicious entity will be banned for the future. This requirement is also related to the first one as it participates on the privacy protection.
- **Exchange of resources**. This requirement is essential for trust negotiation since it emerges from its nature. Trust is established by the gradual exchange of credentials between negotiators and the participating entities should utilise a secure and reliable method for their transfer. The exchange process should be balanced

as the negotiators expose their privacy to each other evenly. Entities can negotiate on their own or they can use a trust negotiation model to do it for them. For online environments, it is suitable to use a secure channel with a strong encryption.

- **Authority**. During trust negotiation, entities may need to be more confident about claims of the others. In case of doubts, an entity may ask a trusted authority to confirm these claims. Depending on the scenario, an entity may let the authority to sign or issue credentials and the other one then queries their verification. This way, a trusted third party is involved in the process, which increases credibility and enhances the building of trust. This requirement can be very utile, however, it is not vital for trust negotiation.
- **Information granularity**. A credential is usually considered as an indivisible atomic unit. However, it may be useful to quantify it for a better control over the disclosed information. The quantification allows one credential to divide into various information levels and the most suitable one will be used in trust negotiation. This way, the entity may protect its security and privacy. Depending on the type of the credential, it can be achieved in two ways: The composite credentials can be simply quantified by dividing into parts, where these parts are combined as needed and then disclosed. Or, the information contained can be smudged in order to provide less precision and confidence. This method is especially suitable for the time and location data.
- **Context sensitivity**. Trust negotiation is always dependent on a given context. Entities negotiate in order to reach a specific goal and the built trust is valid only for this purpose. This is due to the fact that entities have different characteristics and possess various knowledge and skills, so while they can be trusted in one field, not necessarily can be trusted in another one. Additionally, entities may choose a purpose for which their credentials will be disclosed, such as building trust only. The recipient must use them in compliance with the sender. Also, this requirement helps to protect security and privacy.
- **Roles**. In order to achieve the specific goal, entities must agree on their roles. They can be predetermined to play a certain role based on their attributes, i.e., an entity may be required to possess certain features. Only such entity may become trusted. Roles are sensitive to the given context as one entity may play a different role for a different purpose. Also, they may determine the entities' access rights to resources and their privileges during trust negotiation. The RBAC approach utilises it.

The most important requirement is the exchange of resources since it is natural as well as essential for trust negotiation. The other important ones are the access control to resources that allows access of the defined resources to the authorised entities and privacy of resources that aims to protect privacy of the resource owner as much as possible. The context sensitivity criterion is important for establishing trust for a given purpose. The rest of the criteria may be required or helpful for specific scenarios of trust negotiation.

3.3 Strategy

Trust negotiation can use various strategies for the exchange of credentials. The main purpose is not to disclose them carelessly whenever they are demanded, but to protect privacy of the negotiators with focus on building a successful trust relationship. The strategy defines the content, order and priority of the exchanged messages. They should be disclosed in such a way that the required level of trust will be reached efficiently while maintaining the privacy exposure as low as possible. The strategy also specifies how trust is derived based on the obtained credentials. Generally, each negotiator may choose his own preferred negotiation strategy. He can aim to follow his demands, such as quick negotiation results or an extra privacy preservation. However, it is important that the used strategies of both negotiators are compatible in order to avoid conflicts, failures and privacy abuses during trust negotiation. The other way is to choose the same strategy for both of them, which ensures compatibility and consistency. Winsborough et al. [23] presents two strategies designed for a different purpose. They use a simplified approach for access control, where each credential is assigned a locking attribute. Only the unlocked ones can be disclosed during trust negotiation. The *Eager* strategy focuses on building trust quickly and efficiently. As soon as the access policy is satisfied, credentials are marked as unlocked and are immediately disclosed to the other negotiator without being requested. When new credentials are received in exchange, other locked ones become unlocked and are disclosed too. Trust negotiation is terminated, when there are no new credentials received. Even though this strategy leads to establish a trust relationship when possible, it does not care about privacy at all. The *Parsimonious* strategy is more advanced and cares about privacy. Unlocked credentials may be disclosed only if they were requested and the access policy is satisfied. Also, this strategy aims to disclose only their minimal set that meets the requester's demands. Just like in the previous case, as trust is increasing, more locked credentials become unlocked. If a request cannot be satisfied by disclosing unlocked credentials, trust negotiation terminates. Generally, an infinite number of negotiation strategies can be defined. A suitable solution is to define an own strategy for each negotiator that suits his needs and then to agree on them during trust negotiation.

3.4 Access Control and Policies

Policies are essential for trust negotiation. Even when two people negotiate face-to-face, they are using defined policies in their minds. However, for a more advanced scenario, policies have to be explicitly specified in a policy language. In trust negotiation, polices are mainly needed for controlling access to the credentials of negotiators in order to protect their security and privacy. The definition of policies can use various approaches, where the most common ones are the following:

- **Attribute-based Access Control (ABAC).** This paradigm makes use of attributes and their combinations. The basic element is an atomic attribute containing only one value, however, these attributes can be grouped into a composed one. They are of a general type, so they can represent any model or situation. Also, the attributes can be combined by the Boolean logic operators, such as *AND, OR, NOT* and statements, such as *IF, THEN*. These allow to define such rules, where an action is triggered when a condition defined by the combination of attributes is accomplished. For example, IF the requester is authorised AND his account is active THEN allow access to the resource. A relation-based access control is also possible as the attributes can be compared one to each other.
- **Role-based Access Control (RBAC).** This paradigm handles access control based on roles and privileges. The access control decisions are specified by permissions given to particular roles. A user can obtain a permission only if he is assigned a role. Additionally, two conditions must be accomplished: the user must be authorised to be assigned the role he requests and the role must be authorised for applying the particular permission. Roles may form a hierarchic structure, in which the higher-level roles incorporate the permissions from the lower-level ones and can control them.

These access control approaches are implemented in many policy languages in order to define and record policies. Many languages specify their own policy syntax, however, there are also some standard ones defined and used by large companies. The Extensible Application Markup Language (*XAML*) made by Microsoft [15] and the Security Assertion Markup Language (*SAML*) made by the OASIS consortium [11] are general attribute exchange protocols that are also used for the definition of policies. They are based on the Extensible Markup Language (*XML*) that defines a human-readable as well as machine-readable format for encoding documents of various types. For their general usability and portability, these languages are widely used in web services. We will demonstrate and explain the way of how a policy can be specified in the XML language.

Figure 2 depicts an example of the access policy definition for system users in order to permit their access to system resources. The policy forms a tree-like structure, where its hierarchy is determined by the use of multiple encapsulated levels given by pair tags and a superior level consists of all of its descendants. The *<access-policy>* tag represents the root node and contains only the user access policies given by the *<user-access>* tag. Other types of the access policies could be defined subsequently. Two user access policies are specified based on the existing user roles. Each policy creates a connection between the attributes defined in the *<allow-from>* tag with the ones in the *<grant-to>* tag. Also, the blocks and attributes of a policy can form expressions combined by logical operators. In our example, the first policy specifies two roles inside the *<allow-from>* tag that a user can be assigned: "*admin*" and "*developer*". They are encapsulated by the *<or>* tag, which means that this policy grants access for a user playing one or the other role. The second part defined by the *<grant-to>* tag specifies the resource to be accessed. The source code branch "*master*" in the repository "*MICRO*" is defined by the first *<resource>* tag and the

```xml
<?xml version="1.0" encoding="UTF-8"?>
<access-policy>
    <user-access>
        <policy>
            <allow-from>
                <or>
                    <user role="admin"/>
                    <user role="developer"/>
                </or>
            </allow-from>
            <grant-to>
                <and>
                    <resource repository="MICRO" branch="master"/>
                    <resource access="RW"/>
                </and>
            </grant-to>
        </policy>
        <policy>
            <allow-from>
                <user role="tester"/>
            </allow-from>
            <grant-to>
                <resource repository="MICRO" branch="master" access="R"/>
            </grant-to>
        </policy>
    </user-access>
</access-policy>
```

Fig. 2 Access policy example in the XML

read/write access is defined by the second one. They are valid together only as they are encapsulated by the *<and>* tag. The other user access policy is defined alike: A user having the role "*tester*" is allowed to access the same resource, however, only with the read permission.

3.4.1 PlexC: A Policy Language for Exposure Control

PlexC is a policy language designed with trust negotiation in mind and can be well recommended for such use. It brings a new point of view into this field as it introduces a problem of over-exposure [14]. In trust negotiation, when entities exchange credentials, they may provide much more information about themselves than it is necessary to establish a trust relationship and this makes them vulnerable. PlexC identifies this problem as over-exposure that happens quite often during data sharing due to its possible complexity.

Figure 3 depicts a model of the exposure problem. The group *U* represents all the access paths or queries to the confidential data of an entity. They can be dependent on the system, a user context or other past queries. The ideal data sharing model is represented by the group *P**. Performing trust negotiation within this group brings an acceptable exposure. It means that the risk of disclosing credentials is balanced with the benefits, such as establishing a trust relationship.

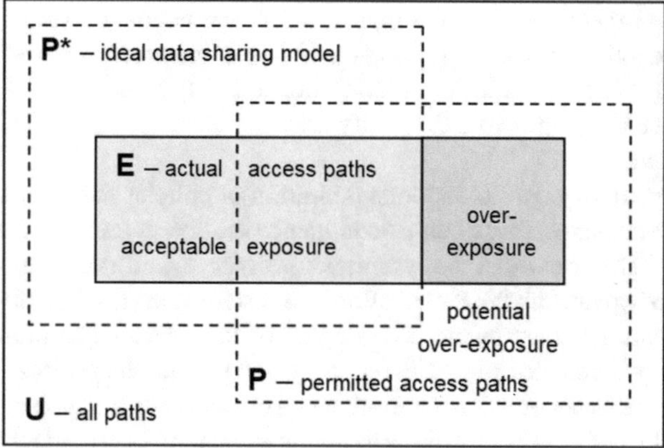

Fig. 3 The exposure control problem

However, the ideal data sharing model is hard to achieve practically due to the complexity of all the included aspects. All the access paths to the resources of an entity, that are permitted by policies, are represented by the group *P*. The actual access paths, that are made to the entity resources, are represented by the group *E*.

From Fig. 3 can be deducted, that:

- $E \cap P^*$ represents the acceptable exposure of an entity.
- $E \cap P \setminus P^*$ represents the over-exposure of an entity.
- $P \setminus P^* \setminus E$ represents the potential over-exposure of an entity.

PlexC supports all of our identified requirements for trust negotiation. They are as follows:

- **Privacy of resources**. Privacy is assured by minimisation of the over-exposure problem. When an entity reaches the desired acceptable exposure area, the risk of disclosures is minimised and privacy of the entity is preserved. PlexC supports querier privacy that serves, e.g., for protecting the identity of the resource requester. It may be useful for anonymous access to resources in large online social networks for entities that, for example, organise protests or share confidential data.
- **Access control to resources**. PlexC can define either simple common policies or more complex ones for information sharing scenarios that serve for the specification of access control. They are referred as exposure-aware, because PlexC handles the access control problem as the exposure problem of an entity.
- **Usage control of resources**. PlexC supports an *exposure control loop* that provides a periodical exposure feedback to entities about the accomplished access paths to their resources and the information of how they are being shared. Then, this feedback can be used to adjust the policies over time in order to avoid the over-exposure of the credentials in the future. This is referred as the exposure polymorphism.

- **Exchange of resources**. PlexC supports disclosure negotiation that is a synonym for trust negotiation. PlexC supports a trust establishment between entities by exchanging credentials and information in open distributed systems. This is very helpful since it is impossible to specify trust relationships between all the entities in the system.
- **Context sensitivity**. PlexC supports definition of policies that may be dependent on different contexts. These contexts influence the access traces to the private data of an entity. They can be of the system or user type, e.g., time, location or activity.
- **Information granularity**. PlexC allows to disclose information about an entity with different accuracy levels. The degree of the provided information can be defined by policies. For example, an entity can provide its position precisely by the GPS coordinates or less accurately by disclosing only a region or a city name. Additionally, definitions of rules are supported that are based on the current time or location of an entity. For example, the access to the resources can be permitted only if the entity is located within a certain region or during a certain period of time. An entity can share its data more or less precisely in order to preserve its privacy.
- **Authority**. PlexC facilitates to delegate decisions about disclosures of resources to a trusted third party. This is mainly useful in large decentralised systems, where the requester and the authoriser may not have established a direct trust relationship. A disclosure decision of an entity may be inspired by a disclosure decision of another trusted entity. The delegation can simplify policies in a hierarchical system.
- **Roles**. PlexC allows the entities to be assigned to one or more roles or groups. Permissions can be given to the whole defined group, which simplifies their management. It is very useful for larger networks.

Due to these characteristics we have chosen PlexC as the most suitable language for the definition of policies for trust negotiation. PlexC mainly aims to preserve privacy of entities while the data sharing model still satisfies their needs. This is especially important for online environments, where the highest number of trust negotiations is happening and the privacy preservation is of great importance.

4 Trust Negotiation Scenarios

Trust is essential for various scenarios in real life. When it was not acquired by a direct experience, trust negotiation can be very convenient. For its versatility, it can be applied in traditional scenarios, where trust is established literally face-to-face, as well as in online environments, where trust is built between total strangers. We will present some examples and demonstrate how trust negotiation can be used.

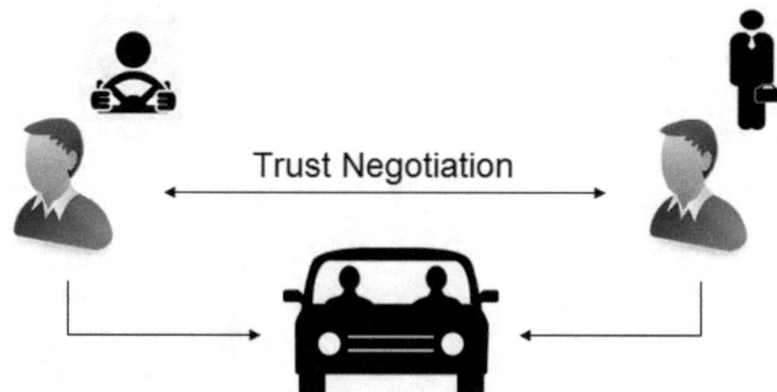

Fig. 4 Trust negotiation simple scenario

4.1 A Simple Scenario

Let us begin with one of the simplest scenarios for trust negotiation. Figure 4 depicts two persons that want to share ride. Their main goal is the same, which is a comfort transport to another city. However, their particular sub-goals differ, but are complementary. The first person is playing a role as a driver that wants to share ride with his car in order to reduce his expenses, whereas the other one does not mind to pay for a faster transport than taking a bus or train. At first, they need to exchange their contact information, such as the e-mail address or phone number to be able to communicate. Then, they need to agree on conditions that will be favourable for both of them, which is basically price for the transport in this case. The driver may request the potential passenger to disclose its home address or working place in order to pick him up and also to gain some confidence about him. The passenger will be probably curious about the driver's profile, so he may ask information about the journey duration or chosen route for gaining confidence. After their interaction, both of them must decide, whether their trust has increased enough. If so, trust negotiation was successful and the participants can proceed to their goal.

This simple scenario does not require any trust model. The persons have been carrying out trust negotiation naturally, maybe even without realising its existence. Still, analogies with a trust model can be seen. At the beginning, the participants want to achieve a goal. Their attributes determine the roles they will play in order to reach it. In their mind, they define policies for accessing credentials, which includes a protection of their privacy and conditions that must be accomplished to be willing to trust. Then, they exchange information in compliance with their policies and acquire confidence about the other side. The financing agreement is important to make a deal, however, it does not have a significant impact on building trust in case of adequate price. In this case, the participants mostly build trust by their social perception and instinct.

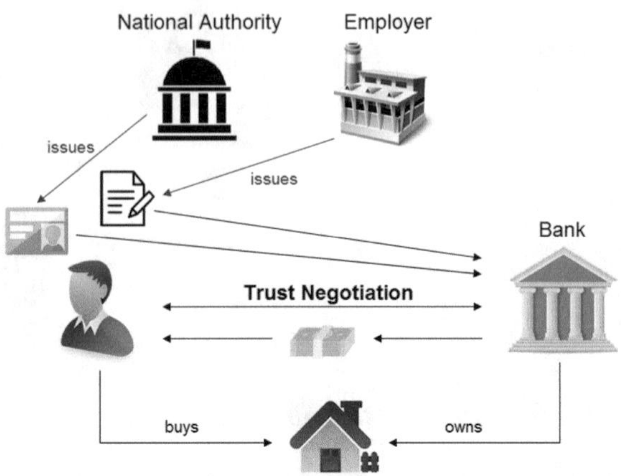

Fig. 5 Trust negotiation advanced scenario

4.2 A More Advanced Scenario

The simple trust negotiation scenario can manage its participants and their implicitly defined goals and policies in their mind. We will demonstrate a more complicated scenario that will require guarantees provided by a third party. Figure 5 depicts a person that wants to take a loan from a bank to buy a property. His goal is obvious, whereas the bank is willing to give money only to a solvent and reliable client in order to profit. The bank must assess potential risks emerging from the loan, while the inputs for its estimation will be obtained from trust negotiation. At first, the client must be identified by disclosing its identity document. The bank trusts this credential as it is an official document issued by a trusted national institution. Providing the loan requires trust of both participants. The client is advantaged as the bank is probably a publicly known institute, has to follow official regulations, is under inspection and has a reputation that helps the client to decide whether to trust. The bank does not know the client and has to build trust with him in order to be confident about his capability to repay the loan on a regular basis.

In this case, trust negotiation is unbalanced and primarily aims to build trust relationship from the bank to the client. The bank has to check his attributes to find out, whether the client is solvent and reliable. Also, the bank has to inspect the attributes of the property to buy, such as its condition, location and price. All of this information will be used by the bank to assess risk of buying the property for its money and for making the final decision, whether the loan can be provided. Specifically, the bank inspects previous loans of the client, if he already borrowed money and the history of his repayment. The client is requested to provide information about his monthly income in form of a signed document issued by his employer. From the bank's point of view, the employer acts as another trusted authority, especially when it is a larger and well known company. Then, the bank matches this information against its policies and makes two decisions: whether the client is qualified to acquire the loan and whether the loan can be provided for the chosen property. If so, the bank establishes

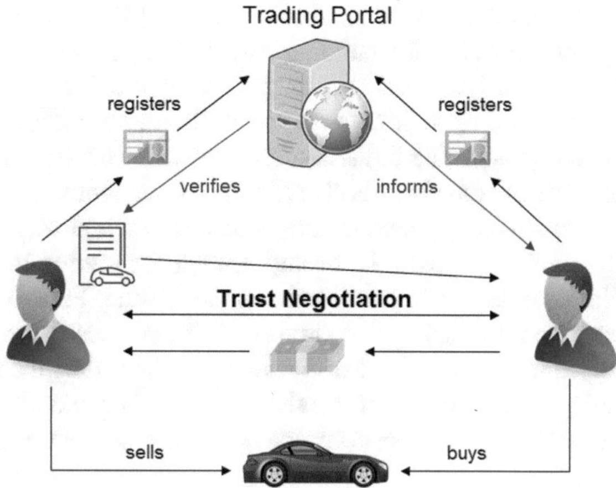

Fig. 6 Trust negotiation online scenario

a trust relationship with the client and lends him money that he will use for buying the property. The bank becomes its official owner, which serves as a guarantee if the client stops repaying. This trust negotiation is suitable for both of them as the client can use the property he would not afford on his own and the bank earns money in a long term.

4.3 An Online Scenario

Trust negotiation is especially suitable for online environments, such as the Internet. Entities join it for various reasons, i.e., for making a business, trading, introducing and chatting. As the Internet is available to billions of people, it presents many challenges in terms of privacy, security and reliability that have to be solved in order to use it with satisfaction. Trust negotiation aims to establish a trust relationship between two unknown and anonymous entities. This is riskier in online environments as the face-to-face contact is missing and entities can forge their identity with less effort. We demonstrate an e-commerce trust negotiation scenario with the use of an online service.

Fig. 6 depicts two persons, a seller and a buyer that want to trade a car using a Web service named Trading Portal. The persons do not know each other, however, they know the service and they trust it. The Trading Portal is publicly known, has a good reputation due to its reliability and is popular for providing support and protection to its users during trading. Based on these known characteristics, newcomers place an implicit trust on the portal without the need of a previous experience and their expectations are mainly positive. At first, the seller and the buyer must register to the portal. They must become its members in order to use its services. During the registration, they have to disclose their personal information, such as name, surname, home address, phone number, e-mail and ID document number. This information serves the portal for building trust to the newcomers, which is actually a specific case of trust negotiation, and for the trading purposes. Also, the newcomers are more aware of their own responsibility in the online trading since they do not act anonymously. The buyer searches the offer of the available cars on the portal and chooses the one that fits his criteria. Subsequently, the seller and the buyer carry out trust negotiation, in which they exchange their chosen personal information in order to establish a trust relationship needed for trading the car. Additionally, the seller discloses its car documentation to the Trading Portal in order to verify it. Then, the buyer is informed by the portal about the verification result, however, no confidential information about the car is revealed to him, thus the portal protects privacy of the seller. In case of positive outcome, the seller and the buyer can meet in person in order to inspect the car physically and eventually trade the car.

5 Trust Negotiation Model

With the identified trust negotiation requirements in mind, we proposed a trust negotiation model that can handle a general scenario of trust negotiation. Figure 7 depicts a UML-based diagram containing the model components and their relations. Let us begin with the component *Negotiator*. This is the main component of the model that needs to establish a trust relationship with another one. In our case, the negotiator can be either the entity itself that wants to negotiate or it can be an entity that carries out trust negotiation on behalf of another one with the necessary delegated rights. The negotiator is characterised by many attributes, from which we focus on the important ones for trust negotiation: *Identity* is essential and identifies the negotiator uniquely. For the purposes of trust negotiation, one needs to know exactly to whom he is building the trust relationship. *Goal* is another one. It describes the reasons why the negotiator decided to build trust. He wants to perform some action, reach an objective, but needs another entity to accomplish it. In order for the other negotiator to be willing to participate too, their goals must be similar or complementary. We say that they have a common goal. *Role* is related to the goal as represents the functions of the negotiators that are relevant for reaching their common goal. Usually, roles are complementary, where one entity needs the other one and vice versa. We depicted a trust relationship coming from the negotiator to itself. The negotiator component can be instantiated and in fact represents two different negotiating entities. In order to communicate, the negotiator uses a *negotiation protocol*. Many different protocols can be defined, however, the negotiators must agree on using the same protocol to be understood one to the other. The protocol defines the format, content and order of the exchanged messages. Similarly, the negotiator uses a *negotiation strategy* that actually drives the whole process of trust negotiation and may heavily influence its results. The strategy represents a plan of how the negotiation will be carried out. It includes aspects like the magnitude of privacy preservation, cautiousness of the negotiator, importance of his goal and many others. As a result, these factors determine

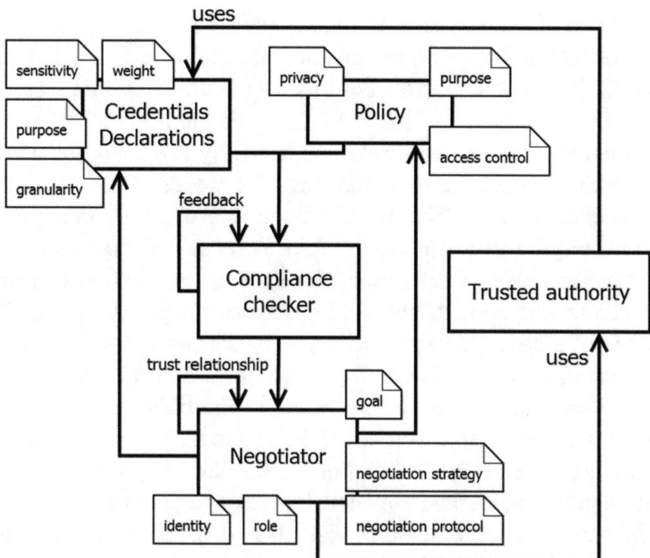

Fig. 7 Trust negotiation model

the credentials to be disclosed, their order and also the way of the overall assessment. The negotiator can define and use his own strategy that fits the best his needs. He does not need to necessarily use the same strategy as the other one, however, the used strategies must be compatible. They must complement, otherwise a deadlock situation may happen, in which both negotiators are mutually waiting for credentials to be disclosed and the negotiation fails.

In order to start with trust negotiation, the negotiators must define their policies that control the exchange process. They are represented by the *Policy* component in our model. We classified them into three types based on their purpose for trust negotiation. The first and the most important ones are the *access control policies (ACP)*. They consist of conditions that must be accomplished in order to permit access to the credentials of the negotiator. The conditions form chains and include various factors that determine the disclosure decisions, such as the identification and authorisation of the other negotiator, the overall built trust level, etc. For many negotiators, especially in online environments, one of these factors is a protection of their privacy. The *privacy policies* focus on this and restrict access to credentials similarly as the ACP. We suggest that each negotiator defines a maximum acceptable privacy exposure level specified in these policies. It should be the worst case of his privacy leakage that it is still tolerable regardless of the trust negotiation results. The last type is expressed by the *purpose policies* that represent the goal-related conditions. They define the exact context for building trust based on the intentions of the negotiator and the way of how trust values are calculated with regard to this context.

The component *Credentials/Declarations* represents the credentials and declarations of the negotiator. They are his private data that usually consist of sensitive information that has to be protected. For trust negotiation, they are essential as trust is built by their exchanging. In this chapter, we define declarations as uncertified claims of an entity that have not been confirmed by a trusted authority. Credentials are declarations that have been confirmed. However, usually we refer to both of them as credentials only. We have identified their features and classified them into the following groups: *sensitivity*, *weight*, *purpose* and *granularity*. Sensitivity characterises credentials by their confidentiality. A sensitivity level attribute can be used for the privacy exposure calculations during trust negotiation. Weight specifies the importance of a credential for building trust. Similarly, a weight attribute may be used for trust calculations, where the weight values of the obtained credentials are summed up and express the overall value of the built trust. Usually, a correlation between sensitivity and weight can be discovered as the more confidential credentials are higher assessed for building trust. Also, weight is related to purpose, because trust negotiation is context-related. Purpose specifies the objective for which the credential should be used. Some credentials may be allowed to be used only for establishing trust, whereas the others may be used after trust negotiation for different purposes. Additionally, purpose defines the goal-related context, within which the credential may be used. The weight value just may vary depending on this context. The last one, granularity, defines the desired magnitude of information disclosure. During trust negotiation, a credential can be split into parts and only its subset or less-detailed information can be disclosed. Granularity is useful for a better control of privacy exposure.

The described credentials/declarations and policy components are in the ownership of the negotiator. In case of a software application, they can be stored as data structures, e.g., in a database. The *Compliance checker* represents a core component for trust negotiation that is delegated by the negotiator to make disclosure decisions. The negotiator demands his own credentials from the compliance checker to be disclosed to the other side. The checker matches these demands against the defined policies and decides, whether the request can be satisfied. Depending on the scenario, the compliance checker can serve only as a suggester of suitable decisions or can be entitled to carry out trust negotiation almost autonomously. Also, the compliance checker can recommend which credentials to disclose next. Based on the used strategy, it may include and consider other important factors, such as the required speed of negotiation or provided feedback. Compliance checkers can be connected together and exchange a feedback about their activities and decisions. This can help to make the negotiation more efficient. They can also share information about the disclosed credentials, whether they were used as intended. Although it may be problematic to find this out, it can be a useful experience for the potential future negotiations. The *Trusted Authority* is the last component in our model. It represents a third party, usually commonly known and trusted by both of the negotiators. The authority can issue new credentials or sign declarations on one side and verify them on the other side. This component is not requisite for trust negotiation, however, for many scenarios the trusted third party is very helpful, such as in online environments.

6 Conclusion

In this chapter, we delved into the concept of trust negotiation that aims to establish a trust relationship between two entities for a specific purpose. We analysed its principal functionality and described the way of how trust is incrementally being built. Then, we summarised our previously identified criteria that are required or helpful for trust negotiation. Based on them, we proposed a trust negotiation model designed to address a general trust negotiation scenario. In this model, negotiators exchange their credentials based on the recommendations of the compliance checker that matches the credential requests against the defined policies. Also, we proposed a few specific trust negotiation scenarios representing various situations. Each focuses on establishing a trust relationship that is required for carrying out the intended actions of its participants. These scenarios represent common trust issues that entities face.

Trust negotiation requires managing access to the resources of entities. We summarised the most common access control approaches and delved into the definition of access control policies. We presented an access policy example in the XML language as well as provided an overview about the PlexC policy language that has been identified as the most comprehensive one for the needs of trust negotiation.

References

1. Becker MY, Sewell P (2004) Cassandra: Distributed access control policies with tunable expressiveness. In: Proceedings of the fifth IEEE international workshop on policies for distributed systems and networks, 2004. POLICY 2004. IEEE, pp 159–168
2. Blaze M, Feigenbaum J, Ioannidis J, Keromytis A (1999) The keynote trust-management system version 2. Tech Rep
3. Blaze M, Feigenbaum J, Lacy J (1996) Decentralized trust management. In: Proceedings 1996 IEEE symposium on security and privacy. IEEE, pp 164–173
4. Blaze M, Feigenbaum J, Strauss M (1998) Compliance checking in the policymaker trust management system. In: International conference on financial cryptography. Springer, pp 254–274
5. Bonatti P, De Coi JL, Olmedilla D, Sauro L (2010) A rule-based trust negotiation system. IEEE Trans Knowl Data Eng 22(11):1507–1520
6. Corritore CL, Kracher B, Wiedenbeck S (2003) On-line trust: concepts, evolving themes, a model. Int J Hum-Comput Stud 58(6):737–758
7. Falcone R, Castelfranchi C (2001) Social trust: a cognitive approach. In: Trust and deception in virtual societies. Springer, pp 55–90 (2001)
8. Gambetta D (1988) Trust: making and breaking cooperative relations
9. Grandison T, Sloman M (2000) A survey of trust in internet applications. IEEE Commun Surv & Tutor 3(4):2–16
10. Herzberg A, Mass Y, Mihaeli J, Naor D, Ravid Y (2000) Access control meets public key infrastructure, or: assigning roles to strangers. In: Proceeding 2000 IEEE symposium on security and privacy. S&P 2000. IEEE, pp. 2–14
11. Hughes J, Maler E (2005) Security assertion markup language (saml) v2. 0 technical overview. In: OASIS SSTC Working Draft sstc-saml-tech-overview-2.0-draft-08, vol 13

12. Jøsang A, Ismail R, Boyd C (2007) A survey of trust and reputation systems for online service provision. Decis Support Syst 43(2):618–644
13. Kolar M, Fernandez-Gago C, Lopez J (2018) Policy languages and their suitability for trust negotiation. In: IFIP annual conference on data and applications security and privacy. Springer, pp 69–84
14. Le Gall YG, Lee AJ, Kapadia A (2012) Plexc: a policy language for exposure control. In: Proceedings of the 17th ACM symposium on access control models and technologies, pp 219–228
15. MacVittie LA (2006) XAML in a Nutshell. O'Reilly Media, Inc.
16. McKnight DH, Chervany NL (1996) The meanings of trust
17. Moyano Lara F et al (2015) Trust engineering framework for software services (2015)
18. Perano KJ, Lee AJ, Winslett M (2007) Trustbuilder2: a reconfigurable framework for trust negotiation. Technical report, Sandia National Laboratories (SNL-CA), Livermore, CA (United States)
19. Seamons KE, Winslett M, Yu T, Smith B, Child E, Jacobson J, Mills H, Yu L (2002) Requirements for policy languages for trust negotiation. In: Proceedings third international workshop on policies for distributed systems and networks. IEEE, pp 68–79
20. Tsiakis T, Sthephanides G (2005) The concept of security and trust in electronic payments. Comput Secur 24(1):10–15
21. Wang YD, Emurian HH (2005) An overview of online trust: concepts, elements, and implications. Comput Hum Behav 21(1):105–125
22. Winsborough WH, Li N (2002) Towards practical automated trust negotiation. In: Proceedings third international workshop on policies for distributed systems and networks. IEEE, pp 92–103
23. Winsborough WH, Seamons KE, Jones VE (2000) Automated trust negotiation. In: Proceedings DARPA information survivability conference and exposition. DISCEX'00, vol 1. IEEE, pp 88–102
24. Winslett M, Yu T, Seamons KE, Hess A, Jacobson J, Jarvis R, Smith B, Yu L (2002) Negotiating trust in the web. IEEE Internet Comput 6(6):30–37

Usage Control for Industrial Control System

Oleksii Osliak, Paolo Mori, and Andrea Saracino

Abstract Cyber-Physical Systems (CPSs) is the key-pillar technology for the implementation of the Industry 4.0 concept. In the industrial sector, a physical entity with internet-enabled capabilities is an example of a CPS. Considering the criticality of the processes controlled by CPS, only authorized entities should have access to those systems under certain conditions. Existing access control approaches implemented in the industrial sector mainly rely on the roles that subjects may have to facilitate the separation of duty concept. However, context information and its mutability over time were out of the scope of implemented access control mechanisms. In this chapter, we investigate the application of the advanced access control paradigm to enable continuous control of Industrial Control Systems (ICS) usage according to context-aware security policies. We provide a framework description along with its implementation in a simulation environment. Finally, the obtained results regarding the system's performance are outlined along with a discussion for potential improvement.

1 Introduction

The *Cyber-Physical Systems* (CPSs) [33] paradigm, which take into account both human and *Information Technology* (IT) as part of the automation processes, has different applications. In the industrial sector, embedded systems were replaced by CPSs to improve manufacturing processes and address business needs. As a part of Critical Infrastructure (CI), *Industrial Control Systems* (ICS) have strict requirements to their security, since the interruption of processes they control may have dramatic consequences. Initially, to satisfy those requirements, ICS were located in

O. Osliak (✉) · P. Mori · A. Saracino
Istituto di Informatica e Telematica, Consiglio Nazionale delle Ricerche, Pisa, Italy
e-mail: oleksii.osliak@iit.cnr.it

P. Mori
e-mail: paolo.mori@iit.cnr.it

A. Saracino
e-mail: andrea.saracino@iit.cnr.it

T. Dimitrakos et al. (eds.), *Collaborative Approaches for Cyber Security in Cyber-Physical Systems*, Advanced Sciences and Technologies for Security Applications,
https://doi.org/10.1007/978-3-031-16088-2_9

physically-secured areas without any connection to external networks, thus making them susceptible primarily to local threats. However, to achieve better performance, the trend toward ICS integration with IT networks became one of the main strategies used by organizations. Such connectivity made those systems vulnerable to cyber-threats [1, 15], since ICS became significantly less isolated, creating a greater need to secure them from external threats.

Therefore, data produced and shared by field devices must be protected from illegal access, its modification, etc. The OPC Foundation[1] has researched the direction of improving the ICS protocols security also considering other requirements, including availability, reliability, etc. Thus, the *OPC Unified Architecture* (OPC-UA) [12] framework proposed in 2006, became a standard technology for enabling secure communication within the industrial domain. It offers a set of security features, starting from authentication with authorization based on roles assigned to users, up to cryptography techniques and secure communication channels in order to address security issues. In fact, access restriction is an important and widely used method to protect resources from illegal access. It is especially important for organizations, which remote control of automated systems within critical infrastructure processes in their business. Nevertheless, role-based authorization itself has limitations since it does not enable control over actions execution and it does not consider the system state changes over the time. Moreover, the security mechanism used by the OPC-UA framework does not consider changes in the system state to use these changes in further decision making [35]. Furthermore, the OPC-UA does not enforce complex policies and do not enable revocation of already granted authorization, when certain conditions do not match requirements defined in those policies. Hence, intentional or unintentionally threats, cannot be prevented by using currently implemented mechanism [22].

In this chapter, we describe a framework that exploits the interconnectivity advantages of the OPC-UA framework and enforces complex policies within the ICS environment. The proposed framework relies on the enhanced instance of *Attribute-based Access Control* (ABAC) [20], called Usage Control (UCON) [21], which enables continuous control over assets usage. Moreover, UCON introduces the key feature aiming at revocation of already granted access by stopping previously authorized actions when the policy conditions are not met anymore. We firstly present and describe the model for authorized access reevaluation within the industrial sector. Secondly, we describe the proposed framework architecture and its components. Then, we present the implementation and deployment in a simulated environment together with relevant policy and describe the policy evaluation workflow. Finally, we report a set of experiments to show the system performance as well as overhead introduced by new security mechanism.

[1] https://opcfoundation.org/.

2 Background

CPSs are complex mechanisms that link the physical world through sensors and actuators, together with the virtual world of information processing. CPSs combine diverse components including software systems, communication technology, sensors and actuators that interact with the real world. Usually, these devices are equipped with wireless and wired communication capacity, which can be configured depending on the application environment. Modern CPSs vary in their characteristics, applications, and levels of operations [7]. Furthermore, the CPSs empowered by Cloud technologies have pushed to new approaches [3, 34] that paved the path to *Industry 4.0*.

2.1 Industrial Control Systems Security

Trend Micro[2] defines *Industrial Control System* (ICS) as a term that is used to describe different types of control systems and associated instrumentation. ICS includes the systems, devices, networks, and controls used for automation of industrial processes.

Following security objectives, in 2006, the OPC Foundation has proposed a new concept of communication architecture for industrial systems called *OPC Unified Architecture* (OPC-UA) [12]. The main purpose of the approach is to provide a secure communication between ICS components which use different protocols. Nowadays, the OPC-UA is recognized by many organizations as the main communication and data modelling technology for Industry 4.0.

The communication in the OPC-UA framework is implemented between the *OPC-UA Servers* and *Clients* [17]. Both OPC-UA servers and clients are the software applications. The first ones provide OPC-UA services, while OPC-UA Clients send messages to servers by using services of the OPC-UA framework. The *AddressSpace* model given by the OPC-UA framework is represented as a set of *Nodes* described through attributes [16, 19]. The OPC-UA Security consists of authentication and authorization, encryption and data integrity via signatures. The authentication uses X.509 certificates exclusively. It relies on the application developer to choose which certificate store the OPC-UA application gets bound to. For instance, it is possible to use the *Public Key Infrastructure* (PKI) of an Active Directory.

In this work, we are focusing on the authentication and authorization security aspects of the OPC-UA framework. The authorization mechanism is based on the *Role-Based Access Control* (RBAC). In fact, the RBAC model allows protecting resources by permitting or denying access to users based on the privileges of the role assigned to real users. As a result, this approach simplifies access management and rights assignment since it allows assigning one single role to multiple users. Moreover, the RBAC model has limitations with respect to other access control approaches. These limitations as well as a description of other access control models described in the following section.

[2] https://www.trendmicro.com/vinfo/us/security/definition/i.

2.2 Access and Usage Control

Controlling access to resources and its restriction is one of the fundamental objectives in cyber-security [5]. Over the past 40+ years many access control models were proposed and implemented. Meanwhile, only a few of them are the most well known and used. Thus, the earliest approach was traditional access controls such as *Mandatory Access Control* [27], *Discretionary Access Control* (DAC) [29], and *Role-Based Access Control* (RBAC) [28]. These models are based on the definition of a set of access control rules called authorizations in the form of subject, object, operation. Authorizations define what entity i.e., a subject, may access what resource i.e., object, and specifying which access action i.e., operations are permitted. However, DAC, MAC, and RBAC are focused on managing access to computational resources and digital information within a closed and trusted security domain. Thus, these models are inherently inadequate to address the new problems of modern applications and contexts. One of the attempts to encompass the benefits of DAC, MAC, and RBAC and cover their limitations, was done by proposing the *Attribute-Based Access Control* (ABAC) [6] model that expresses roles, sensitivity and other properties of users, subjects and objects through *attributes*. Furthermore, ABAC allows expressing environmental attributes, that can affect the evaluation of the request and the result of decision making. However, all these models, including ABAC, protect resources up to the point when the access to perform some particular operation is granted to a subject. Thus, traditional access control models, do not assume any option of control during the access execution. Moreover, in the highly dynamic environments, where attributes can change their values, the ABAC model is insufficient since attribute mutability was out of the scope.

To address these challenges, Sandhu et al. proposed a novel *Usage Control* (UCON) [20, 21] model that covers the limitations of traditional access control models, including ABAC. Since the UCON is an evolution of the ABAC model, it also defines subjects, objects, environment, and operations through attributes. Meanwhile, differently to ABAC, the mutability of attribute values was considered within the UCON model as well as continuity of access decision [26]. In the UCON model, attributes updates are initiated as side effects of accesses and result in a change of an authorization system state [25]. Differently from traditional access control models, UCON enforces security policy before, during and after access execution. Thus, if during the ongoing access attributes changed their values and security policy is not satisfied anymore, the UCON authorization system revokes the granted access and terminates the usage.

Figure 1 depicts the XACML [23] reference model called *Usage Control System* (UCS) [2, 10] that includes seven different components. The main component of the UCS is a *Context Handler* (CH) that acts as a frontend. This component is invoked by the security operations (e.g., subject's request) intercepted by the *Policy Enforcement Point* (PEP). This component is implemented into the controlled system. The UCS can have multiple *Policy Information Points* (PIP) that are invoked by the CH component to retrieve attributes. All attributes are managed by the corre-

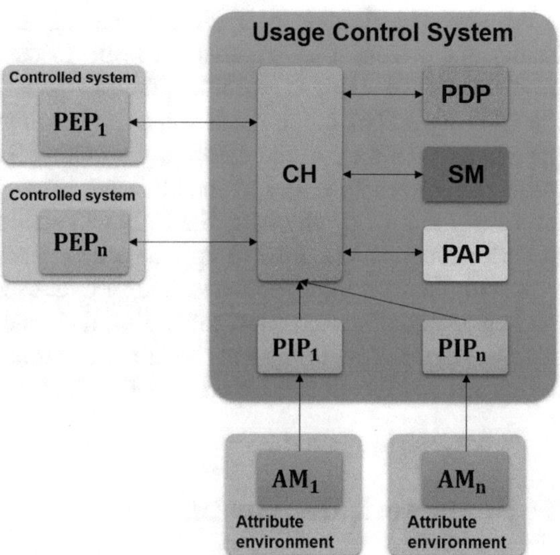

Fig. 1 Usage control system

sponding component called *Attribute Manager* (AM). This component provides an interface used for retrieving attributes and updating their values. Attributes can be required by the *Policy Decision Point* (PDP) in order to evaluate the request according to policies. Usage control policies can be retrieved either from the PEP or *Policy Administration Point* (PAP). The last component is a *Session Manager* (SM) that stores active sessions together with information for policy reevaluation. Finally, since values of attributes can change during the session, the PIP component is responsible for detecting such changes.

There are three different phases of decision process in usage control. Typically, these phases are regulated by the interactions between UCS and PEP components.

- *tryAccess*: belongs to the pre-decision phase. It starts with the TryAccess message from the PEP component to the UCS. PEP creates and send this message when subject requests to execute an access. The tryAccess phase ends when UCS sends the response message to the PEP component. The response could be either "permit" or "deny";
- *startAccess*: belongs to the first part of the ongoing-phase. The startAccess phase begins with the relative message sent to the UCS by the PEP component. The phase finishes after the policy evaluation and when the response has been sent back to the PEP;
- *revokeAccess*: defines the second part of the ongoing-decision phase that is executed whenever an attribute changes its value. The revokeAccess phrase finishes when the policy is evaluated and if a policy violation occurs. Then the UCS sends the RevokeAccess message to the PEP component.

The evaluation of the subject's request begins when the subject tries to execute an action. The PEP component suspends the execution, retrieves attributes related to this request and sends the TryAccess message to UCS. As the next step, the UCS evaluates the request and returns the result to the PEP component. In case if the execution of the action is permitted, the PEP will send the StartAccess message to the UCS right after the moment when the execution of the action started. During the execution of the permitted action, the UCS will evaluate the policy whenever an attribute changes its values. If the attribute value changed and the new value does not satisfy the policy anymore, the UCS will send the RevokeMessage message to the PEP component in order to stop the access session.

To eliminate the imperfections of the OPC-UA security in particular limitations of the implemented RBAC model and protect the system from the attack scenarios [13, 14, 24], where authorized users and/or applications can act with malicious attempts during access execution, we used the UCS aforementioned.

3 Framework Description

In this section we will describe the proposed model, the architecture of the framework that implements it and the deployment in a simulated use case in a smart factory environment.

3.1 The OPC-UCON Model

In this section we propose and describe the Usage Control model for the OPC-UA framework referred as *OPC-UCON* and shown in Fig. 2. The complete description of components and functions is given in the Table 1. The framework combines the OPC-UA standard advantages of interconnectivity and the UCON paradigm that allows monitoring any changes of attribute values and reevaluate the recently given authorization.

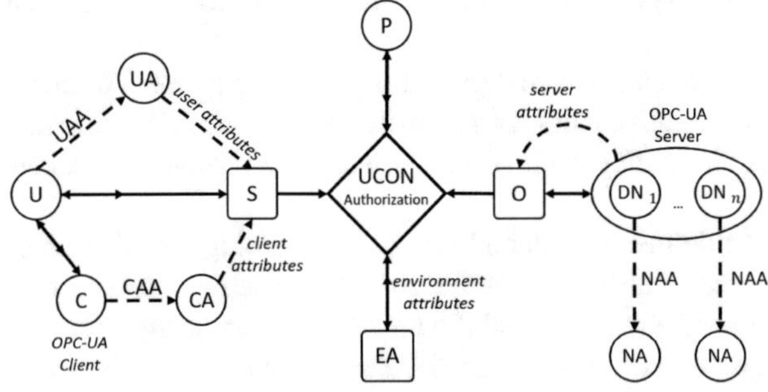

Fig. 2 The OPC usage control model (OPC-UCON)

Table 1 Formal UCON definitions for OPC-UA

Basic Sets and Functions

– U, S, C, Ser are finite sets of Users, Subjects, OPC-UA Clients and Servers respectively;

– O, P and DN are finite sets of Objects, operational Policies and OPC-UA Data Nodes i.e., Node;

– UA, CA, EA, $SerA$ are finite sets of User, Environment, OPC-UA Client and Server attribute functions respectively;

– SA, OA, are finite sets of subject and object attribute functions respectively. The SA set consists of two sub-sets that describe a user and a OPC-UA client i.e., $SA = UA \cup CA$. The OA set is a collection of all DN attributes and $SerA$ i.e., $OA = SerA \cup NA_i, 0 \leq i \leq n$, where n is a number of total NA that describe each Node;

– *UAA*, *CAA*, *NAA* - are functions for user, client and Node attributes assignment;

– attType: $UA = \{set\}$, defines user attributes to be set valued only;

– attType: $SA \cup OA \cup CA = \{set, atomic\}$, defines other attributes to be set or atomic valued;

– For each attribute *att* in $UA \cup CA \cup EA \cup SerA \cup DN$, $Range(att)$ is a finite set of atomic values;

– Each attribute att_{ob} in OA maps $Node$ in DN to attribute values, i.e.,

$$att_{ob} : DN \rightarrow \begin{cases} Range(att_{ob})\ if\ attType(att_{ob}) = atomic \\ 2^{Range(att_{ob})}\ if\ attType(att_{ob}) = set \end{cases}$$

Similarly to the components of the UCON model [20], the OPC-UCON involves finite sets of *Users* (U), *Subjects* (S), *Objects* (O) as well as sets for defining OPC-UA *Clients* (C), OPC-UA *Data Node* (DN) and *Policy* (P) that specify rules and *actions* on data usage within certain conditions. Some of these components have characteristics that are used in access control decision and are expressed through attributes. Hence, user attributes (UA) is a set of user attributes for users and subjects. Additionally, client attributes (CA) are used as additional characteristics for defining subjects. Object attributes (OA) are defined through the OPC-UA server attributes as well as attributes of multiple Nodes. Furthermore, some attributes that characterize environment and belongs to the EA set can be also assigned to the specific Node. However, we distinguish these attributes for the sake of simplicity. Users, OPC-UA clients, server and Node can be assigned attribute values directly for an attribute function *att* from the set of atomic values in the range, denoted by *Range(att)*. Attribute functions in UA set are required to be only set-valued while for OA, CA, and EA sets atomic valued functions are allowed as well. Each attribute function in the UA, denoted by att_u, will map a user to a set of values in a power set of $Range(att_u)$.

In the Table 1 we did not specify the formal description of the authorization functions and other UCON model components (e.g., access operation decisions, obligations) since they remain the same as defined in [21]. Furthermore, we considered user groups and role as additional attributes of users instead of independent entities of the model.

The proposed OPC-UCON model presents a data usage control model for the OPC-UA framework to provide continuous and fine-grained access control to industrial information described through multiple OPC-UA Nodes. The decision result depends on the restrictions described in the policy and current values of subject, object and environmental attributes. Hence, access reevaluation is done whenever the values of attributes change. This fact is critical, especially for the industrial security domain since abuse of information can cause physical damages to systems or even harm the personnel.

3.2 The Envisioned Scenario

To motivate the proposed system, we present here a smart factory environment description, discussing the actual integration of the proposed framework.

We consider a Smart Factory machinery for producing medicine that conjoins different devices for handling the production process within the chemical industry. The production process is similar to the *Tennessee Eastman Process* described in [9]. However, this chapter focuses on the security aspect of the OPC-UA framework.

The control environment of the factory consists of multiple components including a *Programmable Logic Control* (PLC), *Human Machine Interface* (HMI) and *Remote Diagnostics and Maintenance* (RDM) that use OPC-UA infrastructure for communication. Hence, the OPC-UA Server is implemented on the PLC device, while both HMI and RDM acts as the OPC-UA Clients. The PLC device handles the chemical process according to the predefined algorithm by triggering actuators and retrieving relevant data from sensors installed on the production field components. We assume, that the PLC device is similar to Siemens s7-1500[3] since it can support both the Profinet[4] protocol and OPC-UA framework simultaneously. Hence, the PLC device communicates with production field devices by using Profinet protocol. Figure 3 depicts the logical overview of the smart factory infrastructure.

In the smart factory infrastructure that we considered, an operator initiates the process through the HMI by launching two liquid pumps that sway two liquids into the reactor equipped with the heating element and a set of sensors (e.g., temperature, pressure, liquid level gauge). Once the mixing process finished, the PLC device initiates the valve to distill the mixed liquid to the separator. As the final step, the PLC device sends the command to the last two valves to distill liquid and the gas obtained after the separation process. The operator can change the portion of each liquid as well as the max power of the reactor by changing the values of a specific Node.

---

[3] https://opcfoundation.org/products/view/simatic-s7-1500-plc-family.

[4] https://us.profinet.com/technology/profinet/.

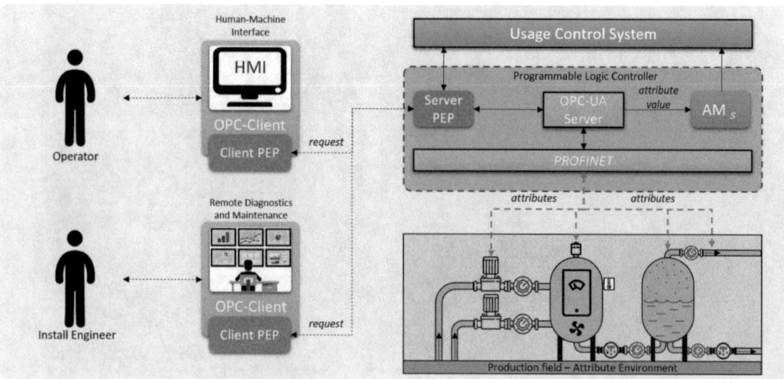

Fig. 3 Smart factory considered infrastructure

3.3 Introducing UCON in the Scenario

As discussed, the OPC-UA framework access management relies on the RBAC mechanism [18] only and credential based authentication. In particular, the authentication process requires users to provide their credentials that include username and password or signed identity certificate. Afterward, specific users can request specific actions on specific information items i.e., OPC-UA Node. With regards to the considered scenario, the machine operator can change the temperature within the reactor by setting up a new value, while the service engineer can set the endpoints for temperature and pressure. However, setting up the new endpoint for pressure can lead to the potential explosion of the reactor and thus harm the personnel or even fatal consequences. In the OPC-UA framework, a security administrator is in charge of the roles assignment. Thus, different roles can be assigned to users depending on their responsibilities in the organization e.g., a working shift supervisor, machine operator, service engineer, etc. Roles can include different access privileges such as *read*, *write*, *edit*, *delete*, *copy*, *execute*, and *modify* a specific Node. Hence, the service engineer can have all the privileges listed above, while the machine operator can only read and write values to the Node. However, the RBAC model does not consider attributes, lacking thus in expressiveness [32]. Furthermore, the mutability of attribute values was out of the scope of the standard RBAC model. Hence, with the RBAC is not possible to model conditions such as working hours, maximum value for the temperature and/or pressure, number of values changes per second, etc.

Figure 4 depicts the view of the integrated UCS into OPC-UA framework. We used multiple Attribute Managers that allow UCS interacting with the physical world by retrieving values from temperature and pressure sensors. The PEP component is implemented both on OPC-UA client and server to support relevant action enforcement depending on the evaluation result of the user's request, thus disconnecting the client from the server or keeping the connection and allowing action execution with relevant notification messages. Such implementation of the PEP component is

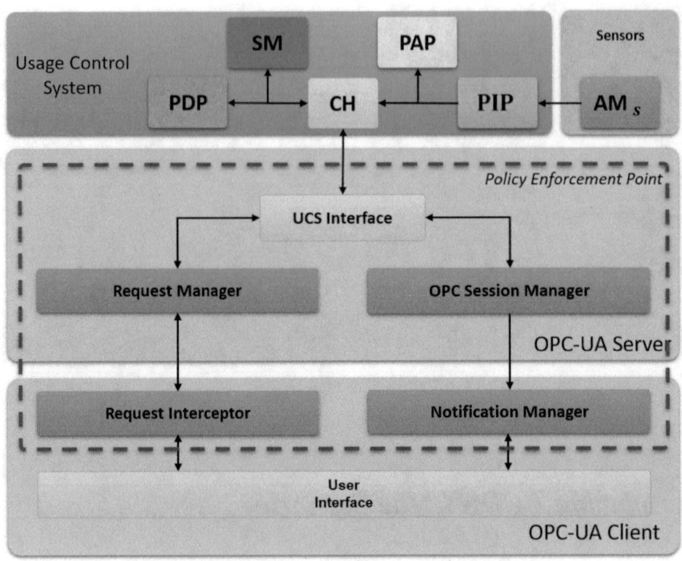

Fig. 4 Integrated UCS into OPC-UA framework

motivated by the necessity of the user identification, registering ongoing OPC-UA communication sessions, communicating with the UCS and if necessary disconnecting clients. To satisfy these requirements, we designed and implemented five sub-components of the PEP described in the following.

The *Request Manager* (RM), *OPC Session Manager* (OPC SM), and *UCS Interface* (UCS IF) sub-components are located on the side of the server, while the OPC-UA client includes *Request Interceptor* (OPC RI) and *Notification Manager* (OPC NM) sub-components of the PEP.

The *Request Interceptor* is in charge of intercepting users data (e.g., username, password) from the OPC-UA client user interface. This sub-components communicates with the Request Manager implemented on the side of the OPC-UA server and with the user interface of the OPC-UA client. The *Request Manager* sub-component is in charge of authentication of clients as well as for creating authorization requests for various actions that users requests to. Requests are written using the XACML standard, which allows defining multiple attributes associated with a certain user including roles, belongings to the specific group, etc. In our case, for authorization requests we considered a *username*, *action*, *Node* and new value. However, it is also possible to define additional attributes e.g., IP address, location, network type, etc. The second sub-component is the *UCS Interface* that interacts with the *Context Handler* component of the UCS. This sub-component is in charge of sending requests from the Request Manager to the UCS and of receiving the evaluation results of requests. Moreover, the UCS Interface delivers messages from the UCS to the OPC Session Manager. The *OPC Session Manager* sub-component is in charge of the access revocation process that might come from the UCS and sending relevant notification messages to the Notification Manager. Additionally, in case of the positive decision i.e., permit, the OPC Session Manager performs the registration of the session. The *Notification Manager* sub-component is used for receiving notifications from the OPC Session Manager and delivering them to the OPC-UA client interface. Thus, whenever the revoke occurred, the Notification Manager will send the message informing users about the policy violation. Also, if the authentication of the user was unsuccessful, the Notification Manager will provide relevant message to the User Interface.

There were no changes done to the UCS and mainly we focused on the integration of the PEP component and its sub-components. Thus, the evaluation process within the UCS described before remain the same.

Figure 5 depicts the workflow of the authorization process including authentication of the client as well as the access revocation caused by the attribute value change. It is worth noting, that the control of operations is done in full respect of protocols. Furthermore, in our work, we apply policies exploiting mechanisms already provided by the OPC-UA framework.

After being authenticated, the user can request a certain action to be executed on a specific Node. Thus, the RI intercepts the requested action together with the Node and a new value to send them to the RM. The RM component accumulates this information together with either ClientID or mapped username and creates the XACML based request. As the next step, the RM sends this request to the UCS IF

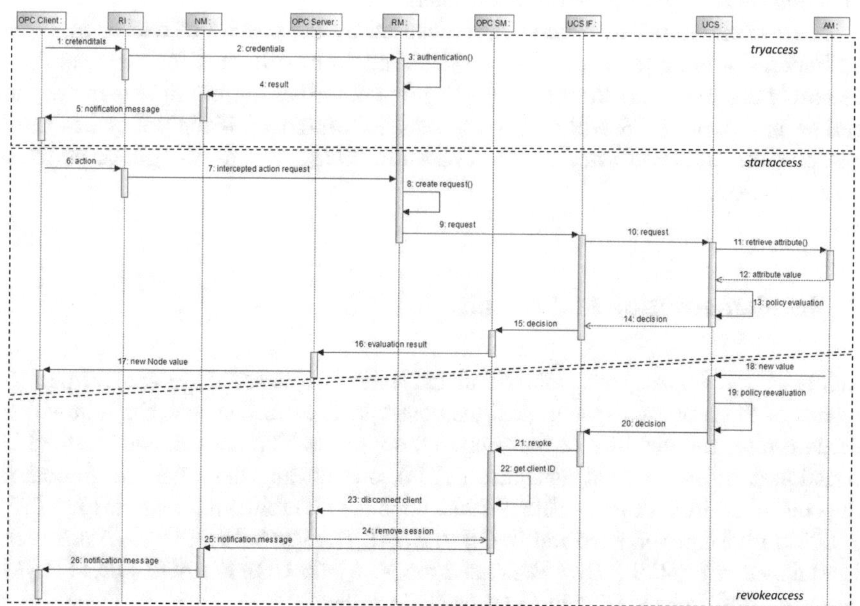

Fig. 5 Authorization and access revoke process workflow

sub-component that interacts with the CH component of the UCS by sending requests and receiving results of the policy evaluation.

Once the UCS receives the request, it retrieves information from one AM or multiple of them as a additional attributes that can be required by the policy. The policy evaluation is done within the UCS by the PDP device that interacts with other components. The complete workflow diagram of the UCS components interaction is described in [11].

Once the request is evaluated the UCS sends the result of the evaluation to the UCS IF. The UCS IF sub-component forwards this message to the OPC SM and if the result of evaluation is positive, the OPC SM sub-component registers the session with the ID number. As the next step, the SM notifies OPC-UA Server about the positive result of the evaluation with the requested action to be performed. As the last step, the OPC Server sends the new value of the Node.

The process of access revocation and disconnection of the client includes 9 steps. Once attributes have changed their values, the Attribute Manager sends new value of each attribute to the Policy Information Point of the UCS. As the second step, the UCS reevaluates the policy with new values. Then, the UCS sends the decision to the UCS IF sub-component of the PEP. As the next step of the revocation process, the UCS IF forwards the decision message to the OPC SM in order to get the Client ID. Once the OPC SM matched the Client ID, it revokes the session and disconnects the client and sends the relevant message to the OPC Server to stop the operation requested by the client. Additionally it removes the session from the list of ongoing sessions and sends the revoke message to the NM to inform the user about the policy violation caused by the attribute value change.

The revocation of the access can be caused by the change of one or multiple attributes. For example, access to the system might be restricted for users between certain hours. Hence, if the attribute value that describes current time is out of the certain limits, the UCS will revoke the session. Moreover, if the policy has been changes, the UCS can force access revocation asking user to re-login or perform another action.

4 Implementation and Results

In this section we will present the simulated testbed exploited to evaluate the performance of the proposed system, and the consequent performance evaluation which demonstrates the viability of the proposed approach. The implemented testbed is constituted by two virtual machines (VMs), one hosting the UCS and the other the controlled PLC devices. The host machine has the following features: Intel i7-7500U with 2 cores enabled and 16 GB of RAM. The VM with OPC-UA framework is equipped with 4GB DDR4 RAM and the VM with UCS has 6GB DDR4 RAM and both of them have Ubuntu 18.04 64-bit installed.

There are multiple open-source implementations of OPC-UA[5] available in different programming languages. For this work, as a first attempt and for the sake of simplicity, we used the client/server version written in Python[6], while the current version of the UCS implemented using Java language. The UCS IF communicates with the CH using the SSL connection. Additionally, we locate the ACL database on the same machine as OPC-UA server. This database is queried by the RM subcomponent to authenticate users.

For this work, we considered a scenario within the industrial process on a simple production field similar to aforementioned. The industrial process is regulated by a single PLC device that communicates with a set of different sensors (e.g., temperature, pressure) and actuators (e.g., heater element, pumps). Real-time values of sensors and actuators are reported as OPC-UA Nodes, where each node specifies the type and the value for the particular component. The PLC device communicates with HMI and RDM through the OPC-UA framework where PLC device is acting as the OPC-UA Server that regulates data usage according to policies, while HMI and RDM act as OPC-UA clients that request various actions.

The policy we consider includes the following attributes: *temperature*, *pressure*, *time*, and the new *value* for the temperature. In our scenario, the user with the *role* of working shift supervisor can set up a new value if it is less than threshold, the request is done within the working hours, the current pressure value is less than the threshold. We consider that the temperature threshold equals 120°C, while for the pressure threshold, we set 5 bar as the max value. Hence, the corresponding policy can be expressed as: "IF *role* is *supervisor* AND *new value* is LESS THAN *120* AND *time* is MORE THAN *08:00* and LESS THAN *18:00* and *pressure* is LESS THAN *5* THEN PERMIT. DENY otherwise". This policy can be easily defined through the *U-XACML* [4] language that is used by the UCS. Since the UCS allows monitoring attribute values change, then if values of aforementioned attributes changed to those that do not satisfy the policy, the UCS will revoke the access and interrupt the connection between OPC-UA server and client. It is worth noting, that we can also consider other attributes such as risk level assigned to a certain user. However, we did not use them for the sake of simplicity.

To validate the proposed approach and measure the overhead caused by the UCS, we run a set of experiments for different number of attributes defined in the policy. Table 2 reports time in milliseconds required by the evaluation process including tryAccess, startAccess and revokeAccess. Hence, it represents the overhead introduced by the UCS.

The second column of Table 2 reports timings required for the tryAccess phase. The third column contains the time required for the startAccess phase. Finally, the fourth column reports the necessary time for the revokeAccess phase process. The actual temperature value change requested by the user is done right after the moment when the startAccess phase finished. Thus, the total time required by the UCS for eval-

[5] https://github.com/open62541/open62541/wiki/List-of-Open-Source-OPC-UA-Implementations.

[6] https://github.com/FreeOpcUa/python-opcua.

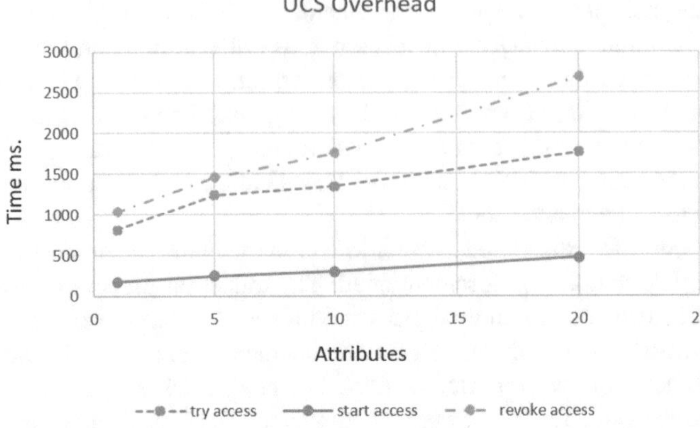

Fig. 6 UCS overhead

Table 2 System performance

Attr. Number	Try access (ms.)	Start access (ms.)	Revoke access (ms.)
1	810	172	1032
5	1238	248	1454
10	1346	297	1752
20	1765	469	2684

uation of the request against the policy with one condition reported via the attribute is less than one second, while for policies with 20 attributes the total evaluation time is less than 1.5 s. However, to the best of our knowledge, such a scenario is unlikely to happen due to the useless complexity of the policy (Fig. 6).

As we expected, the obtained results show that, when the number of attributes increases, the evaluation time increases as well. This fact is caused by the PDP component that adds overhead for evaluating more attributes. Thus, results reported in table were affected by the overhead introduced by the PDP component. Furthermore, implementation of the UCS and OPC-UA on different machines, adds additional overhead caused by the network communication delay. However, considering that the total communication time between the UCS and OPC-UA server is less than 3k milliseconds regardless of the attributes number to be reevaluated by the UCS, the total time for revoking the session is less than the existing threshold of 5k milliseconds, thus making the overhead acceptable for the system.

5 Related Work

In [30], the authors provided an overview of the OPC-UA security and proposed a tool that allows administrators and policy makers to define policies based on the

RBAC model, managing roles and privileges. However, our work aims to improve the security of the OPC-UA by adopting the UCON paradigm to enable continuous control of data usage and, if necessary, revoking the session. In [32] authors proposed the introduction of the ABAC model in the OPC-UA framework. In fact, the security structure with ABAC provides additional security advantages. However, the ABAC model has a lack of control over the mutability of attributes values. Thus, ABAC model introduced in the OP-UA does not allow revoking access. Hence, suck implementation has security limitations comparing to our approach.

In [31] authors presented practical exploitation of Siemens PLC access control vulnerability which can be mitigated by improving management strategy. Furthermore, together with the list of recommendations for improving security, the authors highlighted the importance of fine-grained identity management as well as data usage. The framework proposed in this chapter aims to address issues described by authors by a reevaluation of requests in case of attributes value change. Hence, the framework we proposed will eliminate such issues and make PLC devices, similar to the reported, invulnerable to exploitations reported by authors.

In [8] La Marra et al. proposed a *Usage Control in the Internet of Things* (UCIoT) framework that allows enforcing usage control policies in the IoT architectures. The work is focusing on the implementation of usage control paradigm within the IoT environment. Differently, our work considers an application in a more critical environment, where the solution is integrated seamlessly in the ICS environment, managing operations whose misuse might impose more critical consequences.

6 Conclusions

Industrial systems are becoming increasingly interconnected, among them and with external networks. This evolution enables more flexible operation paradigms, helping stakeholders in handling distributed and delocalized factory architectures. It facilitates remote control and management of the manufacturing processes. However, this openness makes ICS potentially prone to security attacks, which might endanger people's safety and resource integrity. For this reason, introducing access control mechanisms able to enforce security and safety policies is mandatory. In this chapter, we discussed the integration of a framework exploiting the Usage Control paradigm in the OPC-UA framework, already used to secure ICS. The proposed framework strongly enhances the functionalities of the standalone OPC-UA by allowing the definition and enforcement of complex policies considering context information, including the operational environment conditions. The proposed framework implements revocation mechanisms and enables continuous monitoring of mutable attributes. We have provided implementation and deployment in a simulated environment to validate the effectiveness and efficiency of the proposed approach. Further improvement of the proposed framework should consider the complexity of the usage control policies since they directly influence the decision-making time required to evaluate more attributes.

Acknowledgements This contribution was partially supported by the EU H2020 funded project SPARTA, ga n. 830892, EU H2020 founded project NeCS, ga n. 675320 and EU H2020 founded project E-CORRIDOR, ga n. 883135

References

1. Andreeva O, Gordeychik S, Gritsai G, Kochetova O, Potseluevskaya E, Sidorov SI, Timorin AA (2016) Industrial control systems vulnerabilities statistics. Kaspersky Lab, Report
2. Carniani E, D'Arenzo D, Lazouski A, Martinelli F, Mori P (2016) Usage control on cloud systems. Futur Gener Comput Syst 63:37–55
3. Colombo AW, Bangemann T, Karnouskos S, Delsing J, Stluka P, Harrison R, Jammes F, Lastra JL et al (2014) Industrial cloud-based cyber-physical systems. IMC-AESOP Approach 22
4. Colombo M, Lazouski A, Martinelli F, Mori P (2010) A proposal on enhancing xacml with continuous usage control features. In: Grids, P2P and Services Computing. Springer, pp 133–146
5. Easttom C (2019) Computer security fundamentals. Pearson IT certification
6. Jin X, Krishnan R, Sandhu R (2012) A unified attribute-based access control model covering dac, mac and rbac. In: IFIP annual conference on data and applications security and privacy. Springer, pp 41–55
7. Khaitan SK, McCalley JD (2014) Design techniques and applications of cyberphysical systems: a survey. IEEE Syst J 9(2):350–365
8. La Marra A, Martinelli F, Mori P, Saracino A (2017) Implementing usage control in internet of things: a smart home use case. In: 2017 IEEE Trustcom/BigDataSE/ICESS. IEEE, pp 1056–1063
9. Latif H, Shao G, Starly B (2019) Integrating a dynamic simulator and advanced process control using the opc-ua standard. Procedia Manuf 34:813–819
10. Lazouski A, Mancini G, Martinelli F, Mori P (2012) Usage control in cloud systems. In: 2012 international conference for internet technology and secured transactions. IEEE, pp 202–207
11. Lazouski A, Martinelli F, Mori P, Saracino A (2017) Stateful data usage control for android mobile devices. Int J Inf Secur 16(4):345–369
12. Leitner SH, Mahnke W (2006) Opc ua-service-oriented architecture for industrial applications. ABB Corp Res Cent 48:61–66
13. Liang G, Weller SR, Zhao J, Luo F, Dong ZY (2016) The 2015 Ukraine blackout: implications for false data injection attacks. IEEE Trans Power Syst 32(4):3317–3318
14. Libicki M (2015) The cyber war that wasn't. Cyber war in perspective: Russian aggression against Ukraine, pp 49–50
15. Luiijf E (2016) Threats in industrial control systems. In: Cyber-security of SCADA and other industrial control systems. Springer, pp 69–93
16. OPC Foundation (2017) OPC UA Part 3—security model release 1.04 specification. https://opcfoundation.org/developer-tools/specifications-unified-architecture/part-3-address-space-model/
17. OPC Foundation (2018) OPC UA Part 1—address space model release 1.04 specification. https://opcfoundation.org/developer-tools/specifications-unified-architecture/part-1-overview-and-concepts/
18. OPC Foundation (2018) OPC UA Part 2—security model release 1.04 specification. https://opcfoundation.org/developer-tools/specifications-unified-architecture/part-2-security-model/
19. OPC Foundation (2018) OPC UA Part 5—information model release 1.04 specification. https://opcfoundation.org/developer-tools/specifications-unified-architecture/part-5-information-model/

20. Park J, Sandhu R (2002) Towards usage control models: beyond traditional access control. In: Proceedings of the seventh ACM symposium on access control models and technologies. ACM, pp 57–64
21. Park J, Sandhu R (2004) The ucon abc usage control model. ACM Trans Inf Syst Secur (TISSEC) 7(1):128–174
22. Probst CW, Hunker J, Bishop M, Gollmann D (2010) Insider threats in cyber security, vol 49. Springer
23. Rissanen E, Oasis extensible access control markup language (xacml) version 3.0. OASIS Comm Specification 1
24. Robert M, Lee Michael J, Assante TC (2016) Analysis of the cyber attack on the Ukrainian power grid: defense use case. https://ics.sans.org/duc5
25. Sandhu R (2000) Engineering authority and trust in cyberspace: the om-am and rbac way. In: Proceedings of the fifth ACM workshop on Role-based access control. ACM, pp 111–119
26. Sandhu R, Ranganathan K, Zhang X (2006) Secure information sharing enabled by trusted computing and pei models. In: Proceedings of the 2006 ACM symposium on information, computer and communications security. ACM, pp 2–12
27. Sandhu RS (1993) Lattice-based access control models. Computer 26(11):9–19
28. Sandhu RS, Coyne EJ, Feinstein HL, Youman CE (1996) Role-based access control models. Computer 29(2):38–47
29. Sandhu RS, Samarati P (1994) Access control: principle and practice. IEEE Commun Mag 32(9):40–48
30. Schleipen M, Selyansky E, Henssen R, Bischoff T (2015) Multi-level user and role concept for a secure plug-and-work based on opc ua and automationml. In: 2015 IEEE 20th conference on emerging technologies & factory automation (ETFA). IEEE, pp 1–4
31. Wang Y, Liu J, Yang C, Zhou L, Li S, Xu Z (2018) Access control attacks on plc vulnerabilities. J Comput Commun 6(11):311–325
32. Watson V, Sassmannshausen J, Waedt K (2019) Secure granular interoperability with opc ua. In: INFORMATIK 2019: 50 Jahre Gesellschaft für Informatik–Informatik für Gesellschaft (Workshop-Beiträge). Gesellschaft für Informatik eV
33. Wolf WH (2009) Cyber-physical systems. IEEE Comput 42(3):88–89
34. Wu D, Rosen DW, Wang L, Schaefer D (2015) Cloud-based design and manufacturing: a new paradigm in digital manufacturing and design innovation. Comput-Aided Des 59:1–14
35. Xu Y, Yang Y, Li T, Ju J, Wang Q (2017) Review on cyber vulnerabilities of communication protocols in industrial control systems. In: 2017 IEEE conference on energy internet and energy system integration (EI2). IEEE, pp 1–6

UCON+: Comprehensive Model, Architecture and Implementation for Usage Control and Continuous Authorization

Ali Hariri, Amjad Ibrahim, Bithin Alangot, Subhajit Bandopadhyay, Antonio La Marra, Alessandro Rosetti, Hussein Joumaa, and Theo Dimitrakos

Abstract In highly dynamic and distributed computing environments (e.g., Cloud, Internet of Things (IoT), mobile, edge), robust access and usage control of assets is crucial. Since assets can be replicated in various locations on heterogeneous platforms and dynamic networks with unknown or partially authenticated users, the need for a uniform control mechanism is essential. The theory of Usage Control (UCON) is an example of such a mechanism to regulate access and usage of resources based on expressive polices and a loosely-coupled enforcement technology. However, in

A. Hariri (✉) · A. Ibrahim · B. Alangot · S. Bandopadhyay · H. Joumaa · T. Dimitrakos
German Research Center, Huawei Technologies Düsseldorf GmbH, Munich, Germany
e-mail: ali.hariri@huawei.com

A. Ibrahim
e-mail: amjad.ibrahim@huawei.com

B. Alangot
e-mail: bithin.alangot@huawei.com

S. Bandopadhyay
e-mail: subhajit.bandopadhyay@huawei.com

H. Joumaa
e-mail: hussein.joumaa@huawei.com

T. Dimitrakos
e-mail: theo.dimitrakos@huawei.com

A. La Marra · A. Rosetti
Security Forge, Pisa, Italy
e-mail: antonio.lamarra@security-forge.com

A. Rosetti
e-mail: alessandro.rosetti@security-forge.com

A. Hariri · H. Joumaa
Department of Information Engineering and Computer Science, University of Trento, Trento, Italy

S. Bandopadhyay
School of Mathematics, Computer Science and Engineering, University of London, London, UK

T. Dimitrakos
School of Computing, University of Kent, Canterbury, UK

T. Dimitrakos et al. (eds.), *Collaborative Approaches for Cyber Security in Cyber-Physical Systems*, Advanced Sciences and Technologies for Security Applications,
https://doi.org/10.1007/978-3-031-16088-2_10

complex socio-technical systems, concerns about scalability, performance, modularity often arise, and existing UCON models and frameworks cannot meet such requirements. To tackle these concerns, we introduce UCON+, an improvement over existing UCON models, which adds continuous monitoring before granting and after revoking authorizations as well as policy administration and delegation. This chapter aggregates our recent contributions on the conceptual, architectural, and implementation level of UCON+, and provides a comprehensive reference to describe the current state-of-the-art and the novelties of UCON+.

1 Introduction

Access control has been used to protect data privacy and security. Many access control models exist to address different security requirements. They evolved from coarse-grained models such as Mandatory Access Control [13], Discretionary Access Control (DAC) [12] and Role Based Access Control (RBAC) [14] to more flexible and fine-grained models such as Attribute-based Access Control (ABAC) [2]. ABAC provides higher level of granularity by managing access rights using policies that are based on attributes of subjects, resources and the environment. It also incorporates obligations and advice, which are respectively mandatory and optional actions to be completed upon evaluation of policies.

Cyber-physical systems and consumer Internet of Things (IoT) devices, together with an ever-expanding network of sensors and actuators, help interconnect people, appliances as well as Information and Communication Technology (ICT) resources pervading homes, vehicles, healthcare systems and many other aspects of human life. The continuity of interaction combined with the heterogeneity of devices, information, connectivity and automation creates a dynamic environment. Monitoring and controlling permissions, access and usage of information and resources in such environments is essential for maintaining quality, security and safety. However, existing Identity and Access Management (IAM) technologies and models (e.g., ABAC) fail to offer an effective and cost-efficient solution because they assume that access rights do not change during access, so they do not react to situation changes.

For this reason, we introduce a comprehensive model and a novel technology for continuous authorization. Our model builds on the Usage Control (UCON) model [6, 10], which extends ABAC with mutable attributes and continuous control. We designate this technology as UCON+ because it enhances UCON with the following novelties: (i) Extend continuous monitoring to cover interactions before granting and after revoking authorizations; (ii)Enable involving auxiliary evaluators in the decision making process (e.g., trust level evaluation, threat intelligence); (iii) Support policy administration and delegation as well as defining trust authorities and roots of trust; and (iv) Leverage a lightweight policy language for ABAC and extend it with UCON novelties without modifying the syntax and semantics of the language. We also describe Usage Control System Plus (USC+), a modular, scalable and lightweight implementation of UCON+. UCS+ is an robust dynamic authorization technology for embedded systems as well as cloud and Zero-Trust Architectures (ZTAs), benefiting them with:

- Policy-based and code-less behavior at the "brain" of such systems;
- Continuous monitoring of context, environment and data/resource access;
- Intelligent (algorithmic) combination of policies with manageable resolution of conflicts;
- Proactive decisions and interactions to improve user experience, optimize usage of apps and resources, mitigate security and safety risks;
- Automated response to change including revocation of access or restriction of privileges; and
- Ability to advice or prompt for input and involve humans in real-time decision making.

This chapter consolidates our recent advancements on UCON+ with our previous works presented in [1, 4, 5].

2 The UCON+ Model

The UCON model was introduced as a generalization that goes beyond ABAC by adding mutability of attributes and continuity of control. UCON was originally introduced by Park and Sandhu [10] focusing on the convergence of access control and Digital Rights Management (DRM), and extending them with context-based conditions. Unlike preceding models, UCON considers attributes to be mutable, and specifies that policies must be re-evaluated when attribute values change during usage of resources. Lazouski et al. [3, 6] introduced a particularly relevant flavor of UCON that preserves a full ABAC baseline model, complemented with an authorization context, continuous monitoring and policy re-evaluation as well as an implicit temporal state. The temporal state is captured by classifying policy rules as *pre*, *ongoing* and *post*, such that *pre* rules are evaluated before granting authorization, *ongoing* rules are continuously evaluated while authorization is in progress, and *post* rules evaluated upon the end or revocation of authorization. The novelties of UCON enable it to monitor attribute values and re-evaluate policies upon changes in order to guarantee that access rights still hold whilst usage is in progress, or to revoke access if the security policy is no longer satisfied. This makes UCON an excellent baseline for dynamic environments and cyber-physical systems where contextual changes are frequent and authorizations are long-lived.

Continuous control in UCON is limited to the duration of active authorizations only (i.e., only when access has started and is in progress). For this reason, we introduce UCON+, an improvement of the UCON model that extends continuous control to cover interactions before granting and after revoking authorization. This allows UCON+ to cover use-cases that require continuous control and monitoring before and after authorization (e.g., allow subject to improve trust level before granting authorization, ensure that a smart vehicle stops safely upon revoking authorization).

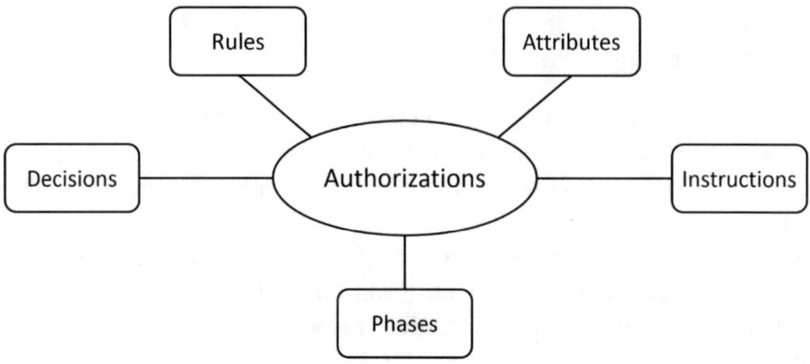

Fig. 1 UCON+ model components

In addition, UCON+ enhances UCON with the ability to involve auxiliary decision factors such as trust level or threat intelligence. UCON+ also extends UCON with the Administration and Delegation Profile (ADP) [11] by incorporating administrative policies that define trust authorities and allows their delegation. In this section, we describe the UCON+ model as well as the policy language used to express UCON+ policies.

2.1 Model Components

The UCON+ model consists of five main components adopted and adapted from UCON [7, 10]. The five components are: Attributes, Phases, Rules, Decisions and Instructions. These components together compose authorizations as shown in Fig. 1 and described as follows.

`Attributes` are properties of the subject (e.g., name), resource (e.g., file type) or environment (e.g., CPU load). Unlike UCON, we consider attributes of all categories (i.e., subject, object and environment) to be mutable, so their values may change while an authorization is in progress. An attribute value may change as a consequence of the authorization itself (e.g., updating metadata of a file upon granting authorization to access it) or due to contextual changes (e.g, subject location changes due to mobility). Since attributes are mutable, the security policy must be re-evaluated whenever an attribute value changes throughout the lifetime of an authorization.

`Phases` refer to the temporal state introduced by UCON and describe the stages of an authorization starting from initial request to enforcement to revocation. Like UCON, UCON+ defines three phases: (1) the *pre* phase indicates that and authorization has been requested but has not been granted yet; (2) the *ongoing* phase denotes that the authorization has been granted and is in progress; and (3) the *post* phase refers to the last stage of the the authorization in which post-usage actions are enforced. In each of these three phases, attribute values and the authorization context are continuously checked in order to promptly react to changes. This enables continuous control in the *pre* and *post* phases in UCON+, which was not enacted in UCON.

`Rules` are functional predicates over attributes that must be evaluated to determine whether an authorization can be granted based on the values of attributes. Rules are classified by phases such that each class is evaluated only when the authorization enters the corresponding phase. For instance, *ongoing* rules must be evaluated only when the authorization has been granted and is in progress. Rules must be re-evaluated whenever an attribute value changes and the authorization decision must be updated accordingly.

`Decisions` are the evaluation outcome of rules. UCON+ extends the classical two-valued decisions (i.e., *Permit*, *Deny*) with two additional values as follows: *Indeterminate* indicates a decision cannot made due to an error or a missing attribute; and *NotApplicable* denotes that the system could not find any rule that matches the authorization request. The decision of an authorization may alter between these values as the context changes throughout the different phases of the authorization as mentioned above.

`Instructions` are mandatory or optional actions defined in security policies and enforced upon evaluation. Instructions include subject obligations, attribute updates and system actions. For instance, a security policy may specify that the metadata of a file must be updated (attribute update). Another example is a security policy specifying that the subject must complete Multi-Factor Authentication (MFA) (subject obligation), and an email must be sent to the sysadmin when the authorization is granted (system action).

2.2 Authorization Session

In traditional access control (e.g., RBAC, ABAC), authorizations are momentary such that only one authorization decision is made per request. UCON introduced the concept of continuous authorization that spans over a period of time, during which the security policy is re-evaluated and the authorization decision is updated accordingly. We designate the context of a single continuous authorization as the *Authorization Session*. An authorization session refers to all events, policy evaluations and contextual information that belong to the lifetime of a single continuous authorization. An authorization session in UCON+ is initiated by and associated with a single *authorization request*. As aforementioned, UCON+ enacts continuous control after the revocation or end of authorization. Thus, an authorization session may last beyond the revocation of authorization to ensure that post-usage actions are completed. Therefore, a UCON+ authorization session starts with a request and ends upon the enforcement of all *post* rules. Attributes that are relevant to a particular authorization session (i.e., required for policy evaluation) are continuously monitored throughout the session, and a re-evaluation of the security policy is triggered whenever an attribute value changes.

2.3 Policy Language

eXtensible Access Control Markup Language (XACML) is the OASIS standard policy language to express ABAC policies [9]. XACML cannot be used to express UCON policies as it does not include phases, which are a main component of UCON and UCON+ as mentioned in Sect. 2.1. To enable the novelties of UCON, Colombo et al. introduced U-XACML [3, 15], an extension of XACML that incorporates UCON concepts. However, U-XACML is incompatible with XACML as it includes some modifications of the standard. For example, U-XACML rules consist of multiple conditions to be evaluated during *pre*, *ongoing* and *post* phases. This change is not compatible with the XACML, which allows only one condition per rule. Thus, U-XACML requires specific modifications to evaluators making its adoption harder. Moreover, XACML and U-XACML are verbose and complex languages, which undermines their readability and efficiency.

To overcome the above drawbacks, UCON+ uses Abbreviated Language For Authorization (ALFA) [8], a pseudocode domain-specific policy language that is in the standardization process by OASIS. ALFA maps directly to XACML without adding any new syntax or semantics, and is used to express ABAC policies. Instead of modifying the ALFA standard, we express UCON+ phases as attributes specified during a usage request, unlike U-XACML which defines phases as conditions within a single rule. ALFA is much less verbose than XACML, which makes it more human readable and shorter in size allowing faster parsing and evaluation. ALFA adheres to the same hierarchy as XACML where decision predicates are expressed in rules that are nested under policies, which in turn are nested under policy sets and/or namespaces. Policies and rules resolve to one of the decisions described in Sect. 2.1 (i.e., *Permit*, *Deny*, *Not Application*, *Indeterminate*) based on their evaluation by the ALFA policy engine. Like XACML, ALFA relies on combining algorithms to resolve conflicts between sibling rules or policies. ALFA also allows the use of functions, such as regular expression, string concatenation and others, which helps to further refine applicability of policies and rules. Listing 10.1 gives an example ALFA policy that limits the action of opening the front door of the house to its owner. The policy adheres to a non-inclusive hierarchy expressed in ALFA syntax as a nested structure of graph parentheses. Thus, each ALFA element is enclosed between parentheses and nested under the parent element.

Listing 10.1 An example ALFA policy

```
policy door {
    target clause Attributes.resource == FRONT_DOOR &&
                  Attributes.action == OPEN
    apply firstApplicable
    rule permitIfOwner {
```

```
6             target clause Attributes.role == "employee"
7             permit
8             on permit {
9                 advice notify {
10                    command = open
11                    resource = door
12                    channel = email
13                }
14            }
15        }
16    }
```

The "target" clause at line 2 determines the applicability of the policy "door" based on the values of the `resource` and `action` attributes, which must be `FRONT_DOOR` and `OPEN` respectively. Line 4 specifies the combining algorithm *firstApplicable* which takes the result of the first applicable rule, in this case the result of *permitIfOwner* defined at line 5. There is only one rule *permitIfOwner* and it is applicable if its *target* matches the value `employee`. If applicable, the decision of this rule evaluates to *permit* and includes an optional instructions (i.e., advice) as expressed in lines 9-14. The instruction assigns the value `open`, `door` and `email` to attributes `command`, `resource` and `channel` respectively. UCON+ also supports administration and delegation ALFA policies, which allow resource owners and administrators to determine who is allowed to issue policies about their resources. This is further described in the following section.

2.4 *Administration and Delegation*

Policy administration controls the types of policies that individuals can create and modify for specific groups of resources. Different classifications such as action type, topic, or some other property are also possible under which policy issuance, revocation or any change can take place. Further operations like delegation are also possible within policy administration, where delegation of authority, or the delegation of the right to issue/revoke policies are possible. For instance, system administrators or resource owners may want to restrict and control who can write policies about their resources. As another example, a resource owner may want to delegate their authority to someone else based on some conditional statements, which could be absence from office or conditions based on time, location etc. The ADP [11] supports such use cases and allows administrators to write policies about other policies forming trees that start with top level policies designated as root of trust.

The truth table mentioned in Table 1 illustrates the outcomes given the *Admissible* and *NotAdmissible* evaluation effects combined with the Usage policy evaluation effects.

Indeed, in some cases, multiple policies may be Admissible and have a Permit/-Deny result. For this reason, a policy reduction algorithm is used, which we discuss in detail in Sect. 4.1. Editing a full-fledged policy with ADP incorporates the steps as mentioned in Listing 10.2.

Table 1 Evaluation outcomes

Usage policy	Administrative policy	Result
Permit	Admissible	Permit
Deny	Admissible	Deny
Permit	NotAdmissible	Find other policy
Deny	NotAdmissible	Find other policy

Listing 10.2 Evaluation and Reduction process

```
<given a root policy defined by authority / organization>
user writes custom policy Pa
synthetic request is crafted from Pa
request is evaluated against Pa
    select administrative policy Padmin for Pa
        for each Pi in Padmin
            create administrative request Ra using R and Pi
            evaluate Ra against Pi
            if deny
                <show modifications>
            else
                <allow policy>
```

In our previous work [1], we introduced the "`PolicyIssuer`" keyword to ALFA, which identifies the author of the policy, and is used to create an administrative request. This enables the use of ALFA to express and issue administrative policies and to support ADP in UCON+.

The policy evaluation flow can be outlined in two steps: reduction and combination. *Reduction* determines whether a usage policy was written by an authorized personnel and whether the evaluation outcome can be considered or not. This is achieved by creating an evaluation tree that has a root-of-trust policy as its root node and the results of applicable policies as its edges, and then finding the branches that can reach the root-of-trust. In the *combination* step, the Policy Decision Point (PDP) combines all valid results using combining algorithms as specified in the policies. The ADP defines a maximum delegation depth to limit the number of administrative policies to be evaluated. Accordingly, the path is discarded if the number of nodes in a path in the reduction graph exceeds the maximum delegation depth.

For both administrative and usage policies, we use the *denyUnlessPermit* combining algorithm due to its deterministic and restrictive nature. This algorithm is restrictive because its default result is always *deny*, unless there is an explicitly applicable permit rule. This also eliminates any indeterministic results like *indeterminate* when an attribute value is missing or a condition is false. Therefore, when an administrative policy evaluates into *indeterminate*, the corresponding branch is discarded from the entire delegation tree.

3 Architecture

This section presents UCS+, an architecture that realizes the UCON+ model presented in the previous section. UCS+ is mainly based on the architecture put forward by the XACML [9] standard and by the UCON framework introduced by Lazouski et al. [6] with several enhancements that we added. We start this section by eliciting the fundamental requirements to be filled by this architecture, and then we describe the components that build up this architecture. Lastly, we illustrate how these components interact to achieve the different required workflows.

3.1 *Requirements and Components Classification*

The main objective of UCS+ is to enable controlling the usage of resources in a uniform way across the broad spectrum of the digital medium, e.g., cloud, edge, terminal, or IoT. The control functionality must accommodate to the mutability of contexts by providing functions to sense the environmental changes (e.g., by monitoring mutable attribute values). In turn, UCS+ must enable methods to grant and revoke decisions of access or usage. To achieve this ambitious goal in a modular, and extensible way, UCS+ comprises three categories of components that are: *fixed*, *programmable*, and *configurable*. Firstly, fixed components are responsible for a specific task regardless of the environment or the application. Secondly, programmable components provide basic behavior and the skeleton of specific functionality. This behavior can be extended to realize any domain-specific requirements. Thirdly, configurable components have defined functionalities that do not change; however, they need to be adapted to cope with specific system properties.

Regardless of their class, all the components communicate using a publish-subscribe model to manage the potential of increasing interactions efficiently. As such, each component registers to events and messages relevant to it. This enables a distributed infrastructure where a component can be spawned in several instances. Thus, UCS+ contains a Message Bus (MSG BUS) to handle our communication paradigm. In the following, we decompose the classes mentioned above into concrete components as illustrated by the Fig. 2.

3.2 *Components Description*

Context Handler (CH) is a *fixed* core component that is responsible for receiving and routing access requests and managing the authorization workflow of such requests. The mentioned MSG BUS is a sub-component that enables CH to achieve this management of the workflow.

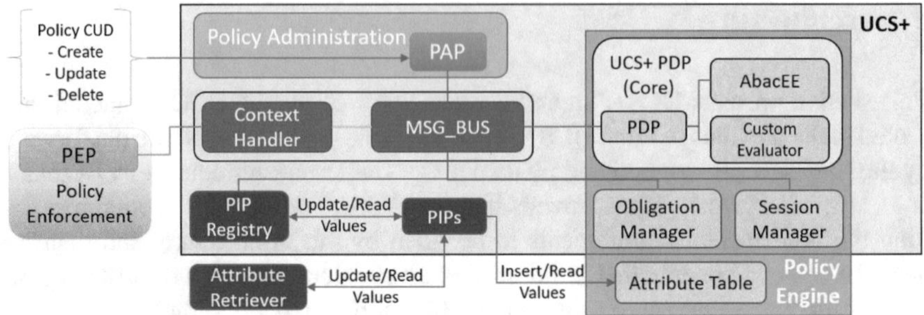

Fig. 2 The architecture of UCS+ illustrating the different component

Policy Administration Point (PAP) is a *configurable* component that acts as a data store of policies, policy sets and namespaces in the system. It offers Application Programming Interface (API) to retrieve, add or update them. This component can be tuned according to different scenarios to persist policies in, e.g., SQL, NO-SQL, or in-memory databases.

Session Manager (SM) is a *configurable* component that is responsible for the continuity of control. Essentially, it is a data structure that stores information about all active sessions. Depending on their authorization workflow, sessions transition among different phases, i.e., *pre*, *ongoing*, and *post*. As such, for a specific session, the first entry in SM is the event following the (Permit) response to the request. Semantically, this means that access has been granted, but the actual usage of the resource still did not start. Later, when the usage of the resource begins, SM changes the status for this session, and UCS+ starts tracking any changes of the mutable attributes. Furthermore, SM keeps a record of the relevant policy for a session, the analyzed request, and a pointer to the relevant attributes. In cases of access revocation, SM updates the session status to Revoked. The session is kept in such a state until all possible obligations are handled, and the access is stopped. Afterwards, the session record is removed from the SM.

Policy Information Point (PIP) is a *programmable* component that defines where to retrieve the values of specific attributes and how to monitor them. PIP is also used to trigger a change in the value of an attribute. A PIP implementation depends on the specific attribute retriever it interacts with; hence, a PIP will be implemented based on the application-specific requirements. To support PIP, we added an additional component, PIP Registry, to manage Policy Information Points (PIPs) and track which PIPs are responsible for which attributes. Attribute Retriever (AR) is another component that supports this process by enabling querying and updating attribute values.

Policy Decision Point (PDP) is a *fixed* component that is responsible for evaluating a request and based on a policy. The result of this evaluation can either be a *PERMIT* (access is authorized), *DENY* (access is denied), *NOT APPLICABLE* (decision cannot be taken due to semantic reasons, e.g., no rules in the policy are matched), or *INDETERMINATE* (decision cannot be taken because either the policy or the request is malformed). To form a decision, the PDP parses the request, resolves the relevant attributes, and then matches them with the values in the policies conditions. To perform the evaluation, PDP may leverage different expression evaluators. The core evaluator in UCS+ is AbacEE to resolver ABAC expressions. UCS+ can also be extended with custom evaluators for particular expressions, such as the one contributed in [4] for Trust Level Evaluation Engine (TLEE) expressions. To handle conflicts, i.e., a single request having several applicable rules, combining algorithms are used to determine the priorities among rules and resolve conflicts.

Obligation Manager (OM) is a *programmable* component that handles a crucial part of the UCON+ model, the obligations. Obligations indicate additional actions that must be performed together with access decision enforcement. Technically, obligations are abstract actions that are not bound to a specific format. For example, obligations can be used to control the values of attributes; in such cases, they are called attribute-update obligations.

Attribute Table (AT) is an auxiliary *configurable* component that we added to manage attributes. It is especially needed in environments with faulty attribute retrievers where there is a possibility that values are not promptly available. While PIP collects the values of the attributes, the AT is responsible for caching these values to handle the periodic polling of values as part of the continuous attribute monitoring. In case the attribute retriever does not include any subscription mechanisms, which would allow the PIP to be notified when an attribute value changes, the viable strategy is to query the attribute retriever periodically. In such a case, AT registers the polling time needed for each attribute according to its criticality.

Lastly, the Policy Enforcement Point (PEP) is the component that integrates the UCS+ with the target application. It is responsible for intercepting usage and access requests within the application and directing them to the UCS+ by interacting with CH. Then, it handles the results of these interactions, e.g., allowing the access, or performing an obligation.

To wrap up this section, we present a sequence diagram of the core workflow of UCS+. Figure 3 illustrates a simplified flow of how UCS+ evaluates.

4 Implementation

UCS+ implementation has been split into three main software components making it extensible, customizable and portable. This structure has been designed to have self-contained codebases and to maximize portability and performance in target environments including IoT, mobile and cloud. The three components are libalfa, ucon-c and ucon-service as described below.

`libalfa` is a library that incorporates the features of the ALFA policy language and encompasses PDP functionalities (e.g., policy parsing and evaluation) and part of the PAP functionalities. It has been entirely developed in C++ with no external dependencies, making it deployable on many different devices and with high performance. libalfa is a self-contained module for ABAC, which enables its use for

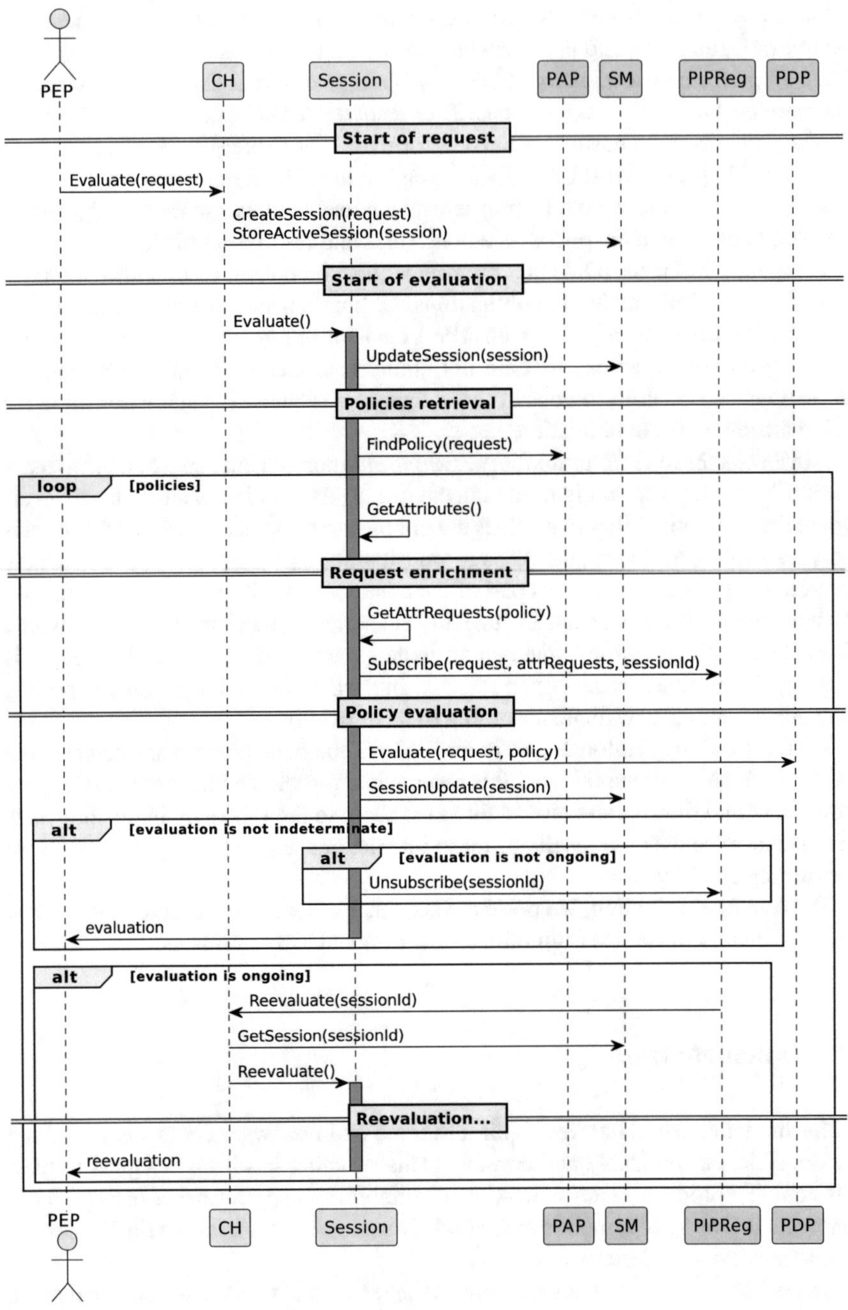

Fig. 3 A sequence diagram of the core UCS+ workflow

scenarios where simple access control is needed without session support or external attributes. libalfa also supports the following two XACML profiles:

1. Multiple Decision Profile (MDP): used in use-cases where access to multiple resources can be requested within a single request.
2. Administration and Delegation Profile (ADP): used to define a root of trust for policies creating a chain of authority to track responsibility and limit attack surface.

`ucon-c` is a library that implements UCON+ novelties such as session-based monitoring and policy re-evaluation according to the architecture described in Sect. 3. It has also been entirely developed in C++ in order to allow an easy integration with libalfa, maximize performance, and enable its deployment on different and low-end devices. ucon-c structure mirrors the block diagram in Fig. 2. Programmable components have been implemented as abstract classes to guide and facilitate their implementation for the requirements of domain-specific use-cases. For instance, PIPs are context-dependent and their behavior varies with the ARs they are connected to, but some of their functionalities are common and do not need to be re-implemented such as caching latest attribute values or enriching requests. ucon-c is *language agnostic* making it adaptable to different policy-languages. Therefore, libalfa can be substituted with other libraries that implement other policy languages such as Open Policy Language (OPA) working with Rego or XACML working with Balana/AuthzForce.

`ucon-service` is a service that exposes an API that enables the use of UCS+ in cloud environments. ucon-service has been developed using Golang due to scalability, performance and resilience requirements. ucon-service exposes a gRPC API that allows managing PIPs and policies as well as initiating usage sessions by external Policy Enforcement Points (PEPs). The service specifically exposes the following gRPC endpoints:

- `Usage` endpoint invoked to start a usage session. It returns a sequence of responses corresponding to the three phases of UCON+ sessions. The endpoint may also respond with an updated usage decision such as revoke, whenever a change occurs and policies are re-evaluated.
- `PAP` endpoint used to create, delete and retrieve policies.
- `PIP` endpoint used to subscribe, unsubscribe and update attributes.

4.1 ADP Reduction Algorithm

According to the ADP, a Usage Policy evaluation result will not be enforced unless evaluation of Administrative Policies result into a permit. Thus, a Usage Policy can be *Admissible* if its enforcement is authorized by all upper layer policies, or *NotAdmissible* if one of the Administrative Policies in the hierarchy cannot be considered. This is expressed in the evaluation outcomes truth table illustrated in Table 1, which shows that whenever Administrative Policies are admissible, a final result is obtained. On the other hand, if one of the Administrative Policies is NotAdmissible, then another policy needs to be found. Indeed, in some cases, multiple policies may be Admissible and have a Permit/Deny result.

A policy reduction algorithm is used to manage inconsistencies and to guide editing of administrative policies in order to avoid a situation where all evaluations end up in a NotAdmissible state. To achieve this, once an administrative policy is added, compliance with the above layers can be checked and suggestions for improvements may be provided. Policy reduction is a process by which authority of a policy associated with an issuer is verified. A reduction tree is built on the basis of the issuer ID of each policy in a policyset. When an access/usage policy is evaluated, the reduction graph is searched for a trusted administrative policy, based on which the combined evaluation effect will be taken for consideration by the PDP. Listing 10.3 describes the reduction algorithm given a policy set and a request. The policies are combined as usual with the defined combining algorithms.

Listing 10.3 Evaluation and Reduction process

```
<given a policy set PS and a request R>
evaluate R against PS
find applicable policies Papp
for each applicable policy Pa in Papp
    if deny
        continue
    if PolicyIssuer is absent
        then combine
    else
        [selection step] : select administrative policy Padmin for Pa
            for each Pi in Padmin
                create administrative request Ra using R and Pi
                evaluate Ra against Pi
                if deny
                    no edge
                else
                    add edge
                    if !policyIssuer
                        potential path is found
                    else
                        go to selection step with Pi as Pa

combine edge results with PS combining algorithm
```

4.2 *Performance*

UCS+ has been designed to work on resource-limited devices such as IoT as well as cloud and high-performance environments. We evaluated the performance of UCS+ in the following different circumstances:

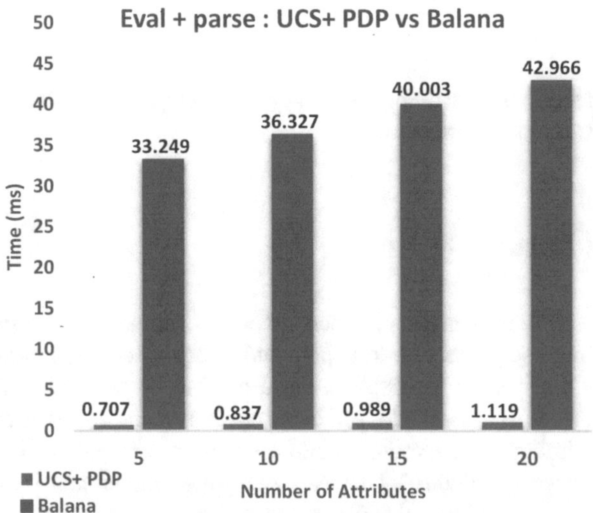

Fig. 4 Performance of UCS+ ALFA Evaluator versus Balana XACML Evaluator

Table 2 Policy length comparison (in KB)

Number of attributes	ALFA	XACML
5	1.1	8.2
10	1.8	16
15	2.6	24
20	3.4	32

- *Cold start*: No attributes are installed in the system. We measured both the end to end time to obtain an evaluation and re-evaluation time, considering the communication penalty and assuming all missing attributes come in the same time.
- *Standard run*: All attributes are already installed.
- *Re-evaluations*: Attributes are supplied one by one at different times in order to evaluate how much time the system takes to perform re-evaluations if attributes are not all immediately supplied.

We also evaluated the performance of the *PDP* to show the time it takes to perform a policy parsing and evaluation compared to Balana[1] XACML evaluator as shown in Fig. 4 and Table 2. We only measured the performance of ABAC part of UCS+ PDP in order to be coherent in our performance evaluation as Balana does not support UCON. The results show that our C++ implementation is more than 40 times faster than Balana.

[1] https://github.com/wso2/balana.

5 Use-Cases

This section describes to example use-cases for leveraging UCON+ in smart home and smart vehicle environments.

5.1 Smart Home

One of the possible use-cases is a smart home environment where it is important to provide safety and protection to people and things since these are connected to the Internet and access to the outside world need to be controlled. For example, multimedia devices used by the kids at home, such as game consoles and smart TV, connect to the Internet and expose kids to a variety of contents that are extremely difficult to manage and control. UCON+ could make significant improvements if combined with the smart home devices and sensors compared to the existing parental controller model where parents have to create a child account for each multimedia device and set limited factory-defined policies. Thus, UCON+ security policies can be specified for safety purposes. For example, if the camera detects that the parents are sleeping or away from home, the child is not allowed to use the smart TV or access the Internet. Moreover, UCON+ obligations enable parents to define specific actions that must be completed before allowing their kids to use such smart devices. For instance, parents can specify an obligation that notifies them when a child attempts to use a smart device, and wait for their approval before allowing the child to use the device. UCON+ supports such cases by monitoring the environment through a variety of devices such as cameras and sensors and using their input as attributes for policy evaluation, allowing dynamic control in the smart home environment.

5.2 Smart Vehicle

Vehicles are real-time applications that are accessed by different entities (e.g., drivers, passengers, applications) in different environments. Thus, they require a dynamic authorization mechanism that can provide continuous usage control to their resources as the context changes. UCON+ can be leveraged to enable such functionality in smart vehicles where access policies are re-evaluated when the location of the car changes as it moves. For instance, the minimum age to drive is 17 in Denmark and 18 in Sweden. Thus, UCON+ policies can specify that a 17-year old driver must be notified when they cross the border from Denmark to Sweden. UCON+ policies may also incorporate obligations that can gracefully stop the vehicle. Car rental is another example where car owners can restrict the use of their vehicles based on rental agreements. A car owner, for example, can designate a specific area in which the driver is permitted to drive. The car owner can also specify obligations that notify them if the driver breaks the conditions of the rental agreement. The owner can also specify a maximum speed or a specific driving period.

6 Conclusions and Future Work

The UCON model was introduced as a generalization of access control to support attribute mutability and continuity of control. Specifically, UCON continuously monitors attributes, then re-evaluates the security policy when an attribute value changes and revokes the authorization if the policy is no longer satisfied. UCON, however, does not enact continuous monitoring before granting 0r after revoking the authorization, which is necessary for proactive and safety-critical systems.

We proposed UCON+ to solve the limitations of UCON, enhance it with policy administration and delegation as well as auxiliary evaluators, and extend it with continuous monitoring that covers pre- and post-authorization interactions. UCON+ uses ALFA as a baseline policy language and solves the problems of U-XACML as described in Sect. 2.3. We presented a reference architecture for UCON+, designated as UCS+, which is adapted from the architecture of UCON. We also described an optimized and lightweight implementation of UCS+, and leveraged it in previous works in different domains such as IoT [4], smart vehicles [5] and data protection [1]. Our implementation incorporates the world's first ALFA evaluator, which also supports policy administration and delegation. We fuse our previous works on UCON+ with our recent progress on the subject and present them in this chapter.

Our future directions aim for formalizing the model and applying formal verification methods to it. We also plan to further develop UCON and UCS+ by capturing a more granular temporal state and adding more phases to the authorization session. Additionally, we intend to validate UCON+ in more domains such as cloud services and identity management systems and evaluate it in both centralized and distributed environments.

Acknowledgements The authors would like to acknowledge the contributions of Yair Diaz and Athanasios Rizos who participated in the research and development of UCON+ at Huawei's Munich research center.

References

1. Bandopadhyay S, Dimitrakos T, Diaz Y, Hariri A, Dilshener T, Marra AL, Rosetti A (2021) DatAPAL: data protection and authorization lifecycle framework. In: 2021 6th South-East Europe design automation, computer engineering, computer networks and social media conference (SEEDA-CECNSM). IEEE
2. Chung, Ferraiolo D, Kuhn D, Schnitzer A, Sandlin K, Miller R, Scarfone K (2019) Guide to attribute based access control (ABAC) definition and considerations. https://tsapps.nist.gov/publication/get_pdf.cfm?pub_id=927500

3. Colombo M, Lazouski A, Martinelli F, Mori P (2009) A proposal on enhancing XACML with continuous usage control features. In: Desprez F, Getov V, Priol T, Yahyapour R (eds) Grids, P2P and services computing [Proceedings of the CoreGRID ERCIM working group workshop on grids, P2P and service computing, 24 Aug 2009, Delft, The Netherlands]. Springer, pp 133–146. https://doi.org/10.1007/978-1-4419-6794-7_11
4. Dimitrakos T, Dilshener T, Kravtsov A, La Marra A, Martinelli F, Rizos A, Rosetti A, Saracino A (2020) Trust aware continuous authorization for zero trust in consumer internet of things. In: 2020 IEEE 19th international conference on trust, security and privacy in computing and communications (TrustCom), pp 1801–1812. https://doi.org/10.1109/TrustCom50675.2020.00247
5. Hariri A, Bandopadhyay S, Rizos A, Dimitrakos T, Crispo B, Rajarajan M (2021) SIUV: a smart car identity management and usage control system based on verifiable credentials. In: IFIP international conference on ICT systems security and privacy protection. Springer, pp 36–50
6. Lazouski A, Martinelli F, Mori P (2012) A prototype for enforcing usage control policies based on XACML. In: International conference on trust, privacy and security in digital business. Springer, pp 79–92
7. Martinelli F, Matteucci I, Mori P, Saracino A (2016) Enforcement of U-XACML history-based usage control policy. In: International workshop on security and trust management. Springer, pp 64–81
8. OASIS (2015) Abbreviated language for authorization version 1.0. https://bit.ly/2UP6Jza
9. OASIS (2017) eXtensible access control markup language (XACML) version 3.0 plus errata 01. http://docs.oasis-open.org/xacml/3.0/xacml-3.0-core-spec-en.html
10. Park J, Sandhu R (2004) The UCONABC usage control model. ACM Trans Inf Syst Secur (TISSEC) 7(1):128–174
11. Rissanen E, Lockhart H, Moses T (2014) XACML v3.0 administration and delegation profile version 1.0. Committee Draft 4. https://docs.oasis-open.org/xacml/3.0/xacml-3.0-administration-v1-spec-en.html
12. Sandhu R, Munawer Q (1998) How to do discretionary access control using roles. In: Proceedings of the third ACM workshop on Role-based access control, pp 47–54
13. Sandhu RS (1993) Lattice-based access control models. Computer 26(11):9–19
14. Sandhu RS, Coyne EJ, Feinstein HL, Youman CE (1996) Role-based access control models. Computer 29(2):38–47
15. U-XACML (2015) XACML with usage control (UCON) novelties. https://bit.ly/3FmeqE6

Printed in the United States
by Baker & Taylor Publisher Services